DEVELOPING SILVERLIGHT® LINE OF BUSINESS APPLICATIONS

Developing Silverlight® Line of Business Applications

with the Enterprise Library Integration Pack for Silverlight™

Microsoft Corporation

ISBN: 978-1-62114-008-5

Contents

PREFACE xiii
- What Is This Guide About? xiii
- What Is Enterprise Library? xiv
- Who Should Read This Guide? xiv
  - If You Are New to Enterprise Library xiv
  - If You Are Already Experienced with Enterprise Library xiv
- How Is This Guide Structured? xv
- Stock Trader V2 Reference Implementation xv
- Who's Who xv

ACKNOWLEDGMENTS xxi
  - The Team Who Brought You This Guide xxi
  - The Silverlight Integration Pack for Microsoft Enterprise Library 5.0 Development Team xxi
  - Advisors xxii

1 **Introduction** 1
- Enterprise Library 1
- Integration Packs 2
- Silverlight Integration Pack 2
- What's in This Release 3
  - Supported Scenarios 3
  - Configuration Support 5
  - Validation 6
  - Caching 6
  - Logging 7
  - Exception Handling 7
  - Interception and Policy Injection 8
- Changes Made to Enterprise Library 5.0 in Optional Update 1 8
- Changes Made to Unity Interception 2.1 9

Installing the Silverlight Integration Pack 10
System Requirements and Prerequisites 10
Installation Using NuGet and Extension Manager 11
Installing NuGet into Visual Studio 13
Adding Enterprise Library Components to Your Application 13
Database Scripts 14
Installing the Configuration Console 14
MSI Installation 15
Performing a Clean Installation 16
Upgrading from Enterprise Library 5.0 16
Upgrading from a Version Earlier than Enterprise Library 5.0 16
Installing the Learning Materials 17
Documentation 17
Installing the Stock Trader V2 Reference Implementation 17
QuickStarts 17
Source Code 17

2 Stock Trader V2 Reference Implementation 19
Limitations 19
The Scenario 19
Problems with the Old Stock Trader 10
Stock Trader V2 Overview 11
Overview of Your Account Position 22
Analyze Market Trends 22
Read Relevant News Articles 23
Place Orders to Buy Or Sell Stock 24
Stock Trader V2 Architecture 27
Modular Development 29

3 Configuration 31
Configuring Enterprise Library 31
Configuration in XAML Files 32
Options for Generating XAML Configuration Files 33
Creating Application Configuration Files Using the Configuration Console 34
Configuration Console with Visual Studio Integration 34
Generating XAML Configuration Using the Configuration Console 36
Generating XAML Using the MSBuild Task 38
Generating XAML Configuration Files Using the Command Line 39
Using Custom Enterprise Library Extensions 40

XAML Translation Considerations 43
Mapping Types between .NET Framework Types and Silverlight 43
Silverlight 4 Types in Configuration Will Also Work in a Silverlight Application 44
Security Considerations for Using XAML Configuration Files 44
Consuming XAML Configuration from a Silverlight Application 45
Embedding the XAML Configuration in the XAP 45
Loading the XAML Configuration from the Server 45
Adding Enterprise Library Configuration to an Already Existing Container 47
Circumvent the Browser Cache When Downloading XAML Configuration Files 47
Configuration in the Stock Trader V2 Reference Implementation 48
Requirements 48
Configuration Architecture 48
More Information 50

4 Diagnostic Logging and Tracing 51
Diagnostic Logging and Tracing with Enterprise Library 51
The Need for Diagnostic Logging and Tracing 52
Detecting (Unknown) Errors 53
Analyzing Problems 53
Analyze Usage Patterns 53
Diagnostic Considerations in a Silverlight Application 53
No Access to Regular File System or Windows Event Log 54
Not All Diagnostic Information is Available in Silverlight 54
Web Service Calls Are Asynchronous 54
The Logging Application Block 55
Supported Scenarios 55
Design of the Logging Application Block 56
Differences between Desktop and Silverlight Versions 59
Remote Service Trace Listener 60
Parameters 61
Logging Service Factory 62
Buffering Messages in Memory 63
Buffering to Isolated Storage 63
Configuring the Logging Endpoint on the Server 63
Isolated Storage Trace Listener 64
Maximum Size 65
Reading Log Messages from Isolated Storage 65

Notification Trace Listener 65
Runtime Changes API 66
Logging and Tracing in the Stock Trader V2 67
Stories for Logging and Tracing 67
IT Operations Should Be Able to View Errors in a Centralized Way 68
Power Users Should Be Able to See Critical Logging Information 69
Users Should Be Able to Increase Their Logging Levels at Run Time 70
Developers Should Be Able to See Detailed Tracing Information 70
Implementing Diagnostic Logging and Tracing in Your Application 71
Using the Remote Service Trace Listener 71
Adding the Log Service to Your Web Server 74
Configuring the Logging Application Block on the Server 75
Configuring the WCF Binding in the Silverlight Application 77
Configure the Logging Application Block to Use the Remote Service Trace Listener 78
Using a LogServiceFactory to Configure the WCF Binding Programmatically 80
Customizing the Log Service by Creating A Derived Class 80
Using the Isolated Storage Trace Listener 81
Reading the Log Messages from Isolated Storage 84
Using the Notification Trace Listener 84
Respond to the Notification Event 86
Changing the Logging Levels in the Application 87
Turning Logging On Or Off 87
Changing the Logging Level for a Trace Source 87
Adding a Trace Listener to a Trace Source 87
Changing the Remote Service Trace Listener 89
Changing the Isolated Storage Trace Listener 89
More Information 90

5 Exception Handling 91
Exception Handling Using the Exception Handling Application Block 91
Exception Handling in a Silverlight Application 92
Global Exception Handler 92
Handling Exceptions in Asynchronous Web Services 93
Introduction 94

Exception Handling Application Block 96
Features 96
Handling Exceptions 96
Defining Exception Handling Policies 96
Exception Handlers 96
Exception Handling in the Stock Trader V2 Reference Implementation 97
Goals for Exception Handling in the Stock Trader V2 97
Unhandled Exceptions 98
Handling Exceptions When Calling Services 100
Returning the Results from a Service 100
Correlating the Start and End of an Asynchronous Operation 104
Handling Exceptions in the Asynchronous Service Agent 105
Displaying Error Messages 107
More Information 108

6 Validation 109
Validation in a Silverlight Application 109
The Validation Application Block 111
Supported Scenarios 111
Defining Validation Rules Using Attributes 112
Comparing WCF RIA Services' Validation to the Validation Application Block 114
Changes to Enterprise Library 5.0 115
Performing Cross-Tier Validation Using the Validation Application Block 116
Sharing Validation Rules Using Linked Files 117
Using Business Entities as Data Transfer Objects in WCF 118
Using Partial Classes Or Conditional Compilation 120
Sharing Validation Rules Using WCF RIA Services 120
Sharing Validation Rules Using Configuration 120
Using the Validation Application Block in a WCF RIA Services Application 121
When WCF RIA Services Triggers Validation 122
When the Validation Application Block Triggers Validation 122
Limitations of the WCF RIA Services Default Code Generator 123
Object Validator Attribute and Object Collection Validator Attribute 124
Performing Server-Side Validation with Rule Sets or Validation Rules Defined in Configuration 125

Using the Validation Application Block in a WCF Data Services Application 127
Using the Metadata Type to Define Validation Rules 127
Performing Server-side Validation 128
Implementing Validation in the Stock Trader 139
Validation with WCF Data Services (The Account Holder Info Module) 130
Account Holder Info Module Scenario 130
Validation with WCF RIA Services (the Orders Module) 132
Orders Module Scenario 132
Sequence for Validation 135
Advanced (View) Model Scenarios 136
Implementing the INotifyPropertyChanged Interface Using the Property Support Class 138
Implementing INotifyDataErrorInfo Using the Error Collection Class 140
Validating a Graph of Objects Using the ModelGraphValidator Class 141
Using the Model Graph Validator with WCF RIA Services 143
Automatic Validation When a Property Changes 144
Displaying Validation Errors That Are Not Bound in the UI154
Multi-Threading Support 145

7 Caching 147
Caching Using the Caching Application Block 147
Caching Considerations in a Silverlight Application 148
Memory Is Not Shared with Other Users 148
Isolated Storage Size Limit 149
Browser Caching 149
Server-Controlled Cache Expiration is Difficult to Implement 149
Silverlight Caching Application Block 150
Comparing to Enterprise Library Caching Application Block 150
System.Runtime.Caching 150
Design of the Silverlight Caching Application Block 151
Object Cache and Memory-Backed Cache 151
Cache Item Policy 152
Cache Capabilities 153
In-Memory Cache 154
Limiting the Size of the In-Memory Cache 154
Isolated Storage Cache 154
Limiting the Size of the Isolated Storage Cache 155

Item Size on Disk 155
JSON Serializer 156
Multiple Instances of the Isolated Storage Cache Running at the Same Time 156
IMPLEMENTING CACHING IN THE STOCK TRADER: THE POSITION MODULE 157
Scenario 157
Design of the Position Module 158
Caching Values to Isolated Storage 159
Proactive Caching 159
Asking the User for More Space in Isolated Storage 161
IMPLEMENTING CACHING IN YOUR APPLICATION 163
Configuring the Caching Application Block 163
Adding an In-Memory Cache Using Configuration 163
Adding an Isolated Storage Cache Using Configuration 165
Using the Object Cache 167
Retrieve the Cache from the Container 167
Adding Items to the Cache 168
Testing If an Item Is in the Cache 168
Using the Removed Callback 169
Using the Updated Callback 170

8 INTERCEPTION 171
What Is Interception? 171
Interception to Handle Cross Cutting Concerns 173
Exception Handling as a Cross Cutting Concern 173
Other Examples of Cross Cutting Concerns 174
Benefits of Applying Interception to Cross Cutting Concerns 174
UNITY INTERCEPTION FOR SILVERLIGHT 175
Unity Interception 175
Instance Interception 175
InterfaceInterceptor 176
Advantages and Limitations of Instance Interception 177
Type Interception 177
Interception Behaviors 179
INTERCEPTION AND POLICY INJECTION 180
Matching Rules 181
Call Handlers 183
Built-in Call Handlers 183

9 Interception in the Stock Trader: The News Module 185
Caching through Interception 185
Catching and Handling Exceptions from the Service Boundary 188

10 Applying Interception 191
Setting Up the Container for Interception 191
Intercepting Objects 191
Intercepting an Object Using an Interface 192
Interception Using Virtual Methods 192
Adding Interception to Already Registered Types 193
Adding Behavior to Intercepted Objects 194
Creating a Custom Interception Behavior 194
Implementing Interfaces Using Interception 196
Applying Policy Injection 197
Setting Up a Type Mapping in Unity for Policy Injection 197
Defining Policies in Code 198

Appendix 199
WCF Data Services Validation in Silverlight QuickStart 199
Using the Metadata Type to Define Validation Rules 200
Performing Validation on the Server 202
Performing Validation on the Client 203
Prerequisites 205
Running the QuickStart 205

Index 207

Preface

What Is This Guide About?

This guide is about the Enterprise Library 5.0 Silverlight Integration Pack. It describes how you can take advantage of the features of Enterprise Library in Microsoft® Silverlight® line-of-business (LOB) application development.

Cross cutting concerns, such as exception handling, diagnostic logging, and validation are just as important to a Silverlight application as they are to any other type of application. However, implementing these cross cutting concerns in Silverlight can be quite a different matter. This guide will explain how you can use Microsoft Enterprise Library's Silverlight Integration Pack to implement these cross cutting concerns with the same level of flexibility and configurability that you have come to know in the desktop version.

Enterprise Library 5.0 ships with a huge set of documentation, ranging from conceptual documentation and hands-on labs, to reference materials. Since many of the concepts and techniques in Enterprise Library 5.0 are also used in the Silverlight Integration Pack, those same topics will not be covered in any depth here. Rather, this guide will complement the **Enterprise Library 5.0 documentation** and provide you with a general understanding of the basic concepts and techniques, then focus on solving Silverlight-specific scenarios using Enterprise Library.

This guide presents a sample application called Stock Trader V2 reference implementation to illustrate the typical challenges you might encounter when building a Silverlight LOB application. In addition, comments from fictional personas, including an experienced Enterprise Library developer named Ed and an IT Professional named Mark, help you understand the rationale behind certain design decisions.

What Is Enterprise Library?

Enterprise Library is a set of reusable application blocks that help developers solve common enterprise software development challenges such as validation, logging, and exception handling. These reusable components are based on practices and techniques that have proven to be very effective in enterprise development.

The Silverlight Integration Pack is an extension that can be used in Silverlight 4.0 and has been tested against Silverlight 5.0 beta. It has been built to take full advantage of Silverlight-specific capabilities, but has been designed so that it can work seamlessly with Enterprise Library 5.0 on the server.

Who Should Read This Guide?

This guide primarily targets software developers and software architects who are building LOB applications in Silverlight. It assumes you have a basic understanding of Silverlight. In certain instances, it will also describe how you can use Enterprise Library to extend technologies such as Windows® Communication Foundation (WCF), WCF RIA Services and WCF Data Services.

If You Are New to Enterprise Library

This guide does not assume that you have experience with Enterprise Library. Each chapter will begin with a basic introduction of the most important concepts from Enterprise Library. In many places, there are links to the **Enterprise Library 5.0 reference documentation** on MSDN® or **the Developer's Guide**, which can help you to gain a deeper understanding of the concepts and techniques discussed.

If You Are Already Experienced with Enterprise Library

Developers and architects who have used Enterprise Library in the past will find that the Enterprise Library Silverlight Integration Pack uses many of the techniques and concepts used in the desktop version.

While this guide provides an introduction to Enterprise Library, the main focus is how you can apply Enterprise Library to Silverlight applications.

How Is This Guide Structured?

You can think of the structure of this guide as a subway that has a main line and then splits into several branches. The main line contains general information that applies to all application blocks, whereas each branch discusses a particular application block dedicated to a cross-cutting concern.

While we encourage you to read this guide from beginning to end, you can also follow only the routes that are pertinent to your work. Within a particular track however, you'll get the best results if you read from beginning to end, because each topic within a track builds on the previous one.

Here is the map:

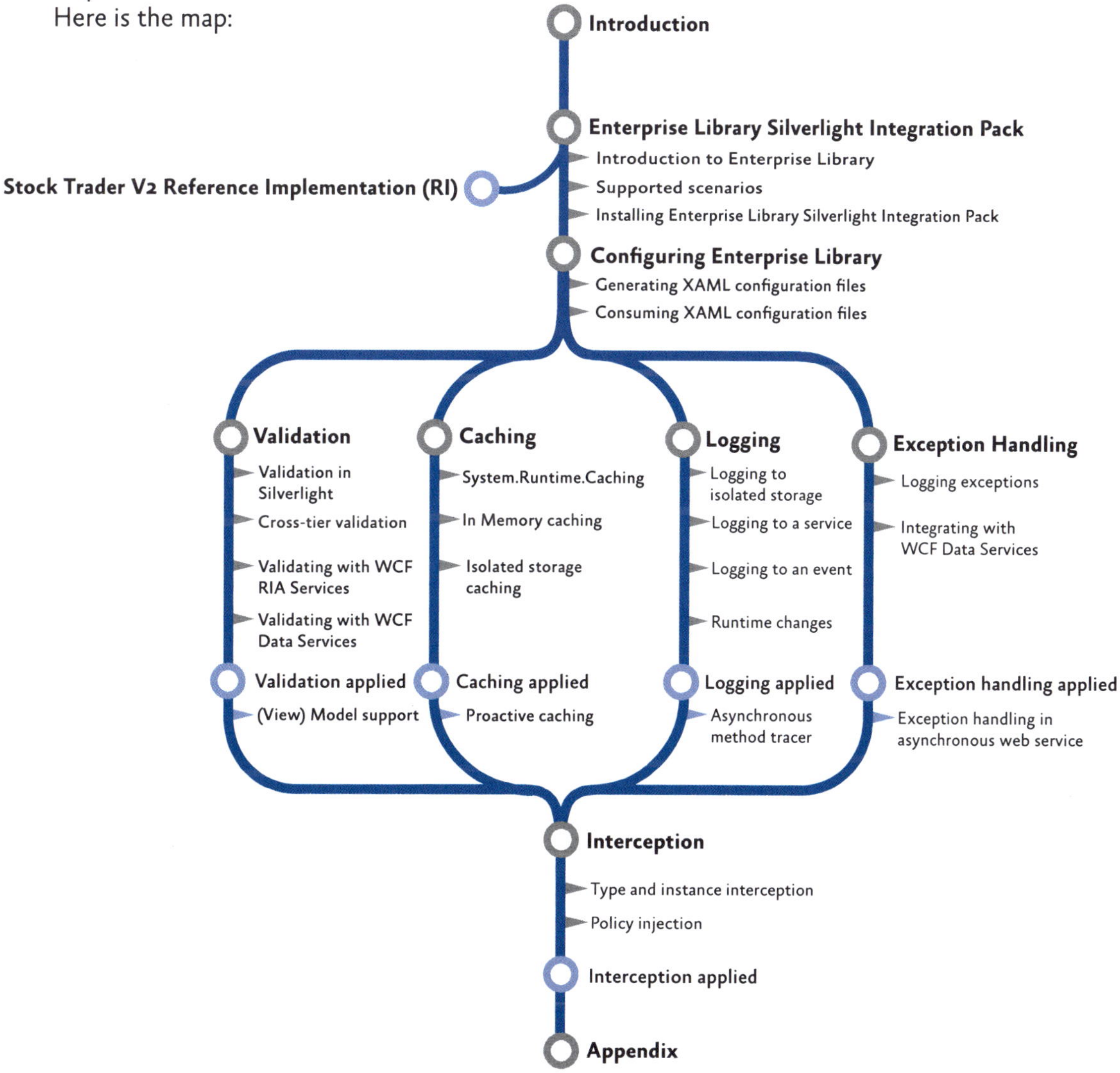

Enterprise Library Silverlight Integration Pack explains the relationship between Enterprise Library 5.0 and the Silverlight Integration Pack. We'll also describe what is in the package, when to use it, and how to install it.

Stock Trader V2 Reference Implementation describes the sample application that ships with the Enterprise Library 5.0 Silverlight Integration Pack.

Configuring Enterprise Library discusses how you can configure the Enterprise Library 5.0 Silverlight Integration Pack. Since Silverlight doesn't support native configuration files, we describe how you can create and use XAML configuration files. Finally, this section discusses how you can share certain configuration information, such as validation rules, between client and server.

Following those sections, the guide splits into four tracks, each covering a particular application block. Each track contains a conceptual introduction, various techniques for applying the block, and a description of how that block was applied in the Stock Trader V2 reference implementation.

The Validation track covers different ways of adding validation logic and business rules to your application. It also discusses several techniques for sharing validation rules between client and server. There is a topic that discusses how you can perform validation if you are using the Model View ViewModel (MVVM) pattern to build your application. This topic also discusses how you can use Enterprise Library Validation with technologies such as WCF RIA Services or WCF Data Services.

The Caching track discusses various techniques for improving the performance of your application, such as caching data to isolated storage. It also describes how you can implement proactive caching in your application.

The Logging track shows how you can add diagnostic logging and tracing to your application. It describes how you can correlate log messages when using asynchronous web services.

Interception is an advanced, but very powerful programming technique that allows you to add new behaviors to an object at run time in a transparent way (for example, applying validation, caching, logging, or exception handling).

Stock Trader V2 Reference Implementation

The Enterprise Library 5.0 Silverlight Integration Pack ships with Stock Trader V2, the sample application mentioned earlier. The application has been developed in Silverlight by the fictitious company, Woodgrove Bank, and will replace the aging Old Stock Trader application.

The chapter called the **Stock Trader V2** explains the scenario and the overall design of the Stock Trader V2. Each chapter that discusses a particular application block also contains a section on how that application block is used by the reference implementation (RI).

Who's Who

When you ask an experienced architect or developer how to solve a particular problem, the most frequent answer you'll get is: "It depends." For each design decision, there are usually several solutions, each with its own advantages and disadvantages. Of course, this also applies to all the design decisions that have been made while developing the Stock Trader V2. While it is valuable to see how you can apply Enterprise Library in a particular situation, it is often just as valuable to know why certain approaches were taken.

In this guide, we'll use the experience of the main contributors to Stock Trader V2 to illustrate why certain solutions were used.

Here are the main contributors to the Stock Trader V2 application:

Ed (The Enterprise Library Expert)

Ed is an experienced software developer and Enterprise Library expert. As a true professional, he is well aware of the common cross cutting concerns that developers face when building LOB applications for the enterprise. In the past, he has built his own libraries to satisfy these concerns, but in the last several years he has used Enterprise Library for most of these applications.

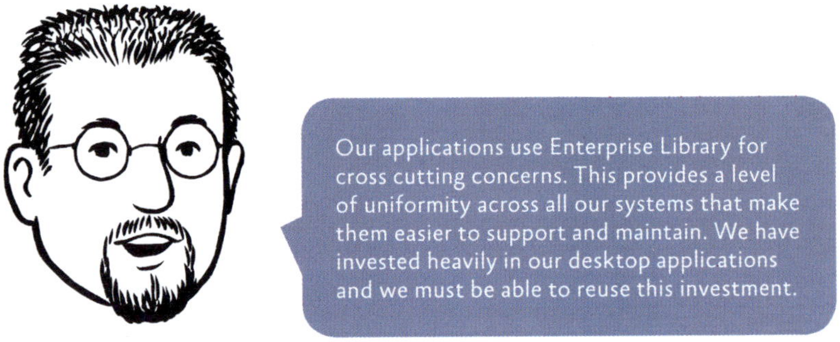

While Ed is highly experienced in UI technologies such as Windows Forms and ASP.NET, he does not have much experience building Silverlight applications.

His main role in the development of Stock Trader V2 was to implement the AccountHolder module. This module was already present in the old Stock Trader application, so Ed has done his best to reuse as much of the original business logic as possible.

Sharon (The Silverlight Developer)

Sharon is an experienced developer of web and Silverlight applications. She's very pragmatic and will always try to choose the best solution for a given problem. When building applications, her primary concerns are to get it to work quickly and get it to work well.

For Stock Trader V2, Sharon has focused mainly on the Orders module. This module had to be rebuilt from scratch. To be as productive as possible, Sharon decided to use the latest technologies, such as WCF RIA Services.

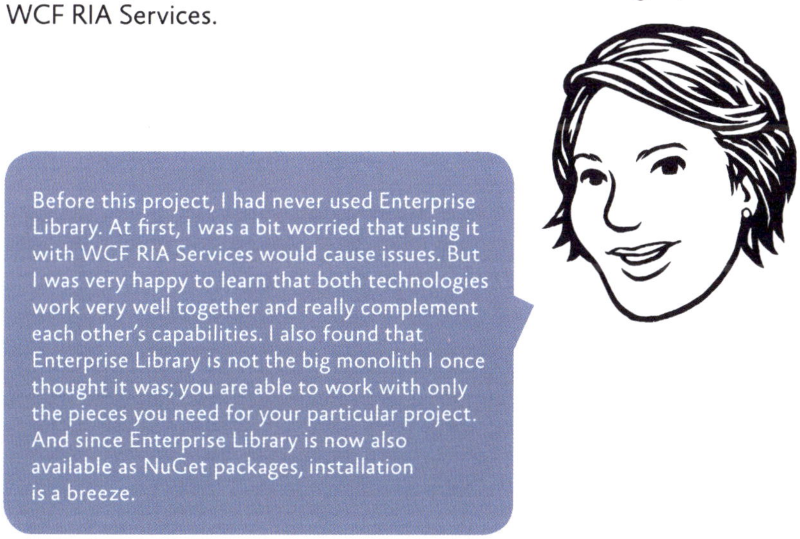

Mark (The IT Pro Responsible for Maintenance)

Mark is an IT professional who is responsible for installing and maintaining software applications. When a problem occurs within a software application, he is usually the first person contacted.

His job is to ensure that Woodgrove's information systems work well and stay up and running. Some of these information systems have been developed in-house, while other systems were developed by third parties but are now being maintained in-house as well.

There is a big difference in quality between some of Woodgrove›s internally developed applications. Some of the older applications are very difficult to maintain because they lack good diagnostic features. Fortunately, some of the more recently built applications have built-in diagnostic functionality.

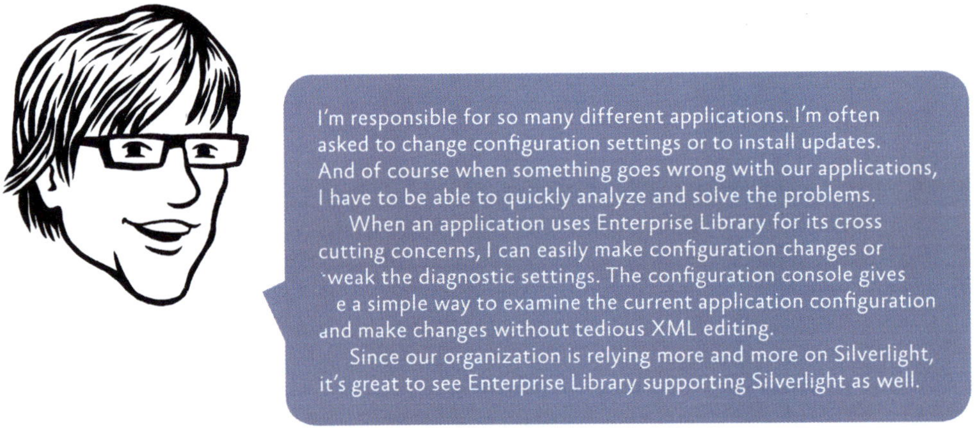

Even though Mark is not a developer, he has been actively involved during the development of Stock Trader V2. He helped to define the stories around design for testability, diagnostic logging and tracing, exception handling, and configuration, and ensured that these were given appropriate priority during the development process.

Acknowledgments

The Team Who Brought You This Guide

Authors	Jérémi Bourgault , Julian Dominguez, Grigori Melnik, Fernando Simonazzi, and Erwin van der Valk
Technical Reviewers	Ward Bell, Colin Bowern, Srdjan Bozovic, Bob Brumfield, David Fowler, Sander Gerz, Phil Haack, Jeff Handley, David Hill, Peter Kuhn, Steve Lasker, Jesse Liberty, Tom Lindeman, Yasser Makram, Ming Man, Paulo Morgado, Erik Mork, Michiel van Otegem, Cigdem Patlak, John Petersen, Daniel Piessens, Erik Renaud, Martín Salías, Jason Shaver, Karl Shifflett, Joe Stagner, François Tanguay, and Shawn Wildermuth
Graphic Artist	Katie Niemer
Book Design	John Hubbard
Editors	RoAnn Corbisier and Nancy Michell

The Silverlight Integration Pack for Microsoft Enterprise Library 5.0 Development Team

Product/Program Management	Grigori Melnik (Microsoft Corporation)
Architecture/ Development	Julian Dominguez and Chris Tavares (Microsoft Corporation), Fernando Simonazzi (Clarius Consulting), and Jérémi Bourgault (nVentive Inc.)
Testing	Mani Krishnaswami and Hanz Zhang (Microsoft Corporation), Srinivasan Vijaya Rengan, Lavanya Selvaraj, Rajarajan Veluchamy, and Sathish Venkateswarlu (Infosys Technologies Ltd.)
User Experience/ Documentation	Erwin van der Valk (Erwin Adrianus Maria van der Valk), RoAnn Corbisier (Microsoft Corporation), Nancy Michell (Content Master Ltd.), and Rick Carr (DCB Software Testing, Inc.)

Release Management Grigori Melnik and Nelly Delgado (Microsoft Corporation), Richard Burte (ChannelCatalyst.com, Inc.), and Jérémi Bourgault (nVentive Inc.)

Advisors

Many people contributed to this release. We are grateful to all of them!

Advisory Council

Ward Bell, Bob Brumfield, Brian Button, David Hill, Peter Kuhn, Jesse Liberty, Erik Mork, Daniel Piessens, Erik Renaud, Karl Shifflett, Joe Stagner, Francois Tanguay, and Shawn Wildermuth

We would like to highlight the exceptional contributions of Daniel Piessens, the most active advisor and Entlib advocate!

Also, we would like to recognize our Microsoft colleagues Jeff Handley for validating our WCF RIA Services Integration story, Steve Lasker for reviewing our Caching Application Block scenarios, David Fowler and Phil Haack for helping with the NuGet packaging and distribution strategy.

Community

Thanks to all who **voted** on the backlog, beta tested our early code drops and provided meaningful feedback. Also, we thank the attendees of the patterns & practices symposia, TechReady, TechEd and NT conferences who provided informal feedback.

1 Introduction

In this chapter, we'll discuss what the Enterprise Library 5.0 Silverlight Integration Pack is and how it relates to Enterprise Library 5.0.

Enterprise Library

Microsoft Enterprise Library is a collection of application blocks intended for use by developers who build complex, enterprise-level applications.

The goals of Enterprise Library are:

- Consistency. All Enterprise Library application blocks feature consistent design patterns and implementation approaches.
- Extensibility. All application blocks include defined extensibility points that allow developers to customize the behavior of the application blocks by adding their own code.
- Ease of use. Enterprise Library offers numerous usability improvements, including a graphical configuration tool, a simpler installation procedure, and clearer and more complete documentation and samples.
- Integration. Enterprise Library application blocks are designed to work well together or individually.

The design of application blocks encapsulates the Microsoft recommended and proven practices for Microsoft® .NET Framework-based application development. These recommended practices are demonstrated in the overall design of Enterprise Library, as well in the context-specific guidelines in the design of individual application blocks and hands-on labs. Software developers can add application blocks to Microsoft® .NET Framework-based applications quickly and easily.

For example, the Logging Application Block helps you implement diagnostic logging by providing a consistent interface that writes log messages to configurable locations, such as a file or a web service.

Integration Packs

Enterprise Library 5.0 is a general-purpose library that can be applied to many different types of applications, such as desktop applications, ASP.NET web sites, and Windows® Communication Foundation (WCF) web services.

While Enterprise Library addresses many development challenges that are common across many different environments, there are also certain development challenges that are unique to a particular environment. For example, when you are building a WCF service, you'll likely want to implement exception shielding. Enterprise Library 5.0 already ships special dlls that only make sense in specific environments, such as exception shielding in a WCF Service or validation support for a Windows Presentation Foundation (WPF) application.

The goal of an Enterprise Library Integration Pack is to take these platform-specific extensions a step further. An integration pack for a specific environment extends Enterprise Library with excellent native support for the most common scenarios in that environment. The first integration pack that ships for Enterprise Library is the Silverlight Integration Pack. Other packs targeting Microsoft Azure, ASP.NET MVC, and Microsoft SharePoint® are envisioned (see the **Enterprise Library roadmap**).

Silverlight Integration Pack

The Enterprise Library 5.0 Silverlight Integration Pack extends Enterprise Library 5.0 to natively support the most important scenarios when building an enterprise-scale Silverlight application. It has been built using Silverlight 4, and has been tested against Silverlight 5.0 beta. Importantly, this guidance is not targeting Windows Phone 7™ Silverlight development.

Since Silverlight uses a custom version of the common language runtime (CLR) and the .NET Framework, it is not possible to use the Enterprise Library 5.0 binaries directly in Silverlight. Therefore, the Silverlight Integration Pack uses a custom-built version of Enterprise Library. This custom-built version shares much of its code with Enterprise Library 5.0.

The custom version does not support all scenarios that the Enterprise Library 5.0 supports; some because they are not possible in Silverlight and others because they simply don't make sense. For example, it is not possible in Silverlight to log directly to the Windows Event Log or to cache directly to a database.

There are certain scenarios and development challenges that are unique to Silverlight development, which are also supported by the Silverlight Integration Pack. For example, the integration pack supports a completely new mechanism for logging messages to a WCF service, solving the problem whereby you cannot log directly to the Windows Event Log.

What's in This Release

- NuGet packages with binaries
  - Enterprise Library 5.0 Optional Update 1
  - Enterprise Library 5.0 Silverlight Integration Pack
  - Unity 2.1 (.NET Framework and Silverlight)
  - Unity 2.1 Interception (.NET Framework and Silverlight)
- NuGet package with the source for Enterprise Library 5.0 Silverlight Integration Pack, which includes the source for the Optional Update 1 as well as Unity 2.1
- Microsoft Windows Installers (MSI)
  - Enterprise Library 5.0 Optional Update 1
  - Enterprise Library 5.0 Silverlight Integration Pack
  - Unity 2.1
  - Unity 2.1 for Silverlight
- Configuration console extension package (VSIX) for Microsoft Visual Studio® development system
- Source code and symbols
- Documentation
- Stock Trader V2 reference implementation
- QuickStarts

Supported Scenarios

In this section, we'll have a closer look at the scenarios that are supported. The following table presents an overview of the support that Enterprise Library 5.0 provides as compared to the Enterprise Library 5.0 Silverlight Integration Pack.

Block	Subject	Supported by Enterprise Library Optional Update 1	Supported by Enterprise Library 5.0 Silverlight Integration Pack
Configuration	Application configuration files	Yes	No
	XAML configuration files	No	Yes
Enterprise Library Core	Performance counters	Yes	No
Validation Application Block	Core Validation Application Block	Yes	Yes
	ASP.NET extensions	Yes	No
	WPF extensions	Yes	No
	WCF extensions	Yes	No
	Windows Forms extensions	Yes	No
Caching Application Block	Core Caching Application Block	Enterprise Library 5.0 implementation	New Silverlight implementation
	Caching to database	Yes	No
	Isolated storage	Yes	Yes
Exception Handling Application Block	Core Exception Handling Application Block	Yes	No
	WCF extensions	Yes	No
Logging Application Block	Core Logging Application Block	Yes	Yes
	Log formatters	Yes	No
	Log to database	Yes	No
	Log to Windows Event Log	Yes	No
	Log to disk	Yes	No
	Log to Windows message queue	Yes	No
	Log to WMI	Yes	No
	Log to email message	Yes	No
	Log to event	No	Yes
	Log to service	No	Yes
	Log to isolated storage	No	Yes
Policy Injection	Core Policy Injection Application Block	Yes	Yes
	Static facade	Yes	No

Unity Interception Application Block	Core Unity Interception	Yes	Yes
	Validation handler	Yes	Yes
	Logging handler	Yes	Yes
	Exception handler	Yes	Yes
	Caching handler	Yes	Yes
Security Application Block		Yes	No
Cryptography Application Block		Yes	No
Data Access Application Block		Yes	No

Configuration Support

The ability to control Enterprise Library through configuration has always been an important feature for Enterprise Library. For example, you can define in configuration where you'd like certain log messages to go, or how to handle certain exception types.

Silverlight doesn't have native support for application configuration files, such as app.config or web.config. To overcome that limitation, the Enterprise Library 5.0 Silverlight Integration Pack allows you to store configuration information in XAML configuration files. You can choose to either embed a XAML configuration file in the Silverlight XAP file or read it from an external location, such as a web server.

You can still use the Enterprise Library configuration console to configure the Enterprise Library 5.0 Silverlight Integration Pack, just as you've done with Enterprise Library 5.0. The configuration console still creates application configuration files, but you'll have several options to generate the XAML configuration files to be consumed from Silverlight:

- **A wizard in the configuration console** that allows you to generate the XAML configuration file directly from the configuration console.
- **A command-line executable** that helps you generate the XAML configuration file via an automated process.
- **An MSBuild task** that helps you generate the XAML configuration in a build process.
- To read more about the configuration support in the Enterprise Library 5.0 Silverlight Integration Pack, please read the chapter on Configuration.

Validation

When you are building an LOB application in Silverlight, you'll likely need some form of data validation support. The Enterprise Library Validation Application Block has been completely ported to Silverlight.

In the chapter on **Validation** in Silverlight, we'll discuss the following scenarios:

- **Sharing validation rules between client and server.** To provide a good user experience, you'll likely want to run as many validation rules as possible on the Silverlight client. However, since Silverlight is a client technology, you must execute all validation rules on the server, because the validation logic on the client can be circumvented. We'll show several techniques for sharing validation rules, such as using Visual Studio linked files or using WCF RIA Services.
- **Validation in a WCF RIA Services application.** WCF RIA Services has built-in support for validation and sharing validation rules between client and server. In this chapter, we'll discuss how you can extend this validation support with certain scenarios that are only provided by Enterprise Library, such as rule sets or validation rules defined in configuration.
- **Cross-tier validation in a WCF Data Services application.** We'll also discuss how you can share validation rules between client and server if you are consuming WCF Data Services in Silverlight.

Caching

A common technique for improving the performance of applications involves caching data. The Enterprise Library 5.0 Silverlight Integration Pack contains an all-new implementation of the Caching Application Block. This new implementation is not based on the Enterprise Library 5.0 Caching Application Block, but is inspired by the **System.Runtime.Caching** namespace, which was introduced in .NET Framework 4.0, but is not yet available in Silverlight.

The Silverlight Caching Application Block supports the following scenarios:

- **In-Memory Caching.** This allows you to store data in memory, with several options for specifying when cached items should expire.
- **Isolated Storage Caching.** This allows you to store data in isolated storage so it is available when the application restarts.

To read more about the new Caching Application Block in the Enterprise Library 5.0 Silverlight Integration Pack, please read the **Caching** chapter.

Logging

Diagnostic logging and tracing is important for any enterprise application. Most of the core of the Logging Block has been converted. You can express in configuration which trace listeners you would like your log messages sent to, just as you can in Enterprise Library 5.0. The Enterprise Library 5.0 trace log providers cannot be used in Silverlight because of the Silverlight security sandbox. Therefore, the Silverlight Logging Application Block ships with its own set of trace listeners.

The Silverlight Logging Application Block supports trace listeners that allow you to perform the following actions:

- **Logging to a remote WCF Service.** The Silverlight Logging Application Block allows you to handle log messages in a centralized way, by logging to a WCF Service. On the server, these log messages are handled by the Enterprise Library 5.0 Logging Application Block. You can send the log messages as they are logged, or in a batch, and buffer them in memory or in isolated storage.
- **Logging to isolated storage.** If you wish to write log information to a persistent store in your Silverlight application, you can also write your log messages to isolated storage.
- **Logging to an event.** The Silverlight Logging Application Block can also raise an event every time a certain log message is received. You can use this to build a custom trace log view in your application, for example.

The Enterprise Library 5.0 Silverlight Integration Pack also supports the use of tracers. A tracer allows you to correlate log messages that are part of the same logical operation. We'll show you how you can apply tracers to asynchronous operations.

To read more about the Logging Application Block in the Enterprise Library 5.0 Silverlight Integration Pack, please read the **Diagnostic Logging and Tracing** chapter.

Exception Handling

The Exception Handling Application Block has been fully ported to Silverlight. With it, you can define how you'd like to handle exceptions by defining exception handling policies.

Exception handling in synchronous method calls is fairly straightforward. However, because all web service calls in Silverlight are asynchronous, we're also showing how you can apply exception handling for asynchronous operations.

To read more about exception handling in a Silverlight application using the Exception Handling Application Block, please read the **Exception Handling** chapter.

Interception and Policy Injection

Interception is an advanced programming technique that provides a non-intrusive way to intercept calls to objects and allows you execute custom code before or after the call. You can use this, for example, to transparently add caching, logging, exception handling, validation, or custom logic to certain objects.

Unity Interception has also been fully ported to Silverlight, with the exception of the **TransparentProxyInterceptor**. Policy injection is also supported through the Unity Interception interface. The chapter on Interception describes in more detail what interception and policy injection are and how you can apply them in your application.

Changes Made to Enterprise Library 5.0 in Optional Update 1

Optional Update 1 contains minor changes made to Enterprise Library 5.0 to support certain Silverlight-specific scenarios. The changes include:

- **Version numbers.** The version number for all Enterprise Library assemblies has been updated to: 5.0.505.0. Even though only a few assemblies have actually been changed, they have all been given a higher version number to prevent version mismatch issues.
- **Validation attributes.** To support the WCF RIA Services code generation, all validation attributes from the Validation Application Block had to be updated to include public read-only properties for each constructor parameter. For the RangeValidatorAttribute and the DateTimeRangeValidator attribute, the LowerBound and UpperBound properties have been changed to type Object.
- **Validation result.** The ToString method is now overridden and will return the message of the validation result. This makes it easier to bind the validation result to a visual element.
- **Validation configuration objects.** To support the XAML markup, the constructor parameters for the validation configuration objects have been removed and converted to properties.
- **Enterprise Library validation extension for Unity fix.**
- **File configuration source with relative paths fix.**

- **Profiles added to configuration console.** The configuration console now supports profiles. If you use the Silverlight profile, only the blocks and options available for Silverlight development are visible. If you use the desktop profile, only the blocks and options that are available for the desktop are visible. You can also choose to run without a profile so that all blocks and options are visible.
- Minor usability fixes to configuration console.
- **Export to XAML wizard added in configuration console.** A new configuration console wizard, called the Export To XAML wizard is shipped with the Enterprise Library Silverlight Integration Pack. This wizard exports the configuration file to XAML, so that the configuration can be consumed from a Silverlight application.

Changes Made to Unity Interception 2.1

The following problems have been fixed in Unity Interception 2.1:

- Hidden Property Causes AmbiguousMatchException in InterceptionExtension.
- Unity Interception: Interface interception problem when using VS 2010.
- The ParameterCollection type in the Unity Interception Extension has a new method to query if a parameter with a given name exists in the collection.
- Using InterfaceInterceptor on a class that contains a method with a generic type that is constrained by another generic type.
- Unity InterfaceInterceptor is not wrapping a resolved derived class that contains no members, even when its base class does.
- Unity VirtualMethodInterceptor incorrectly overrides methods with generic constraints.

Installing the Silverlight Integration Pack

There are several ways to install the Enterprise Library Silverlight Integration Pack:

- **Installation of binaries through NuGet.**
  NuGet is a package manager that makes it much easier to add and configure library packages to your projects. The binaries for Enterprise Library 5.0 and the Enterprise Library Silverlight Integration Pack are available for installation through NuGet. You can use the public NuGet repository or use a local NuGet repository.
- **Install configuration console through Extension Manager.**
  Enterprise Library 5.0 also ships with a Visual Studio Extension, that allows you to launch the configuration console from within Visual Studio. You can install this extension though the Extension Manager in Visual Studio. Alternatively, you can download the VSIX package and install it directly.
    - **Full installation using Microsoft Windows Installer.**
      There are two .msi files that will install all the binaries, the configuration console, and the other support tools needed to use the Enterprise Library 5.0 Silverlight Integration Pack.
    - **Installation from source.** Download the source (either through CodePlex or NuGet) and build it.
- These installation options are described further in this chapter.

System Requirements and Prerequisites

To run an application that was built using the Enterprise Library 5.0 Silverlight Integration Pack, you will need the following prerequisites:

- Silverlight 4.0 or Silverlight 5.0 (Beta).
- A Silverlight 4.0-supported browser and platform. The supported browsers and platforms are described on this page: http://www.microsoft.com/silverlight/faq.

For Remote Logging Service of the Logging Application Block, the following are also required:

- Microsoft .NET Framework 3.5 with Service Pack 1 or Microsoft .NET Framework 4.0.
- Logging Application Block of Microsoft Enterprise Library 5.0 Optional Update 1.

For WCF RIA Services Validation Integration or WCF Data Services Integration, the following are required:

- Microsoft .NET Framework 4.0.
- Validation Application Block of Microsoft Enterprise Library 5.0 Optional Update 1.

These are the minimum system requirements for Enterprise Library 5.0 Optional Update 1:

- Supported architectures: x86, x64.
- Operating system: Microsoft Windows® 7 Professional, Enterprise or Ultimate; Windows Server® 2003 R2; Windows Server 2008 with Service Pack 2; Windows Server 2008 R2; Windows Vista® with Service Pack 2; or Windows XP with Service Pack 3.
- Microsoft .NET Framework 3.5 with Service Pack 1 or Microsoft .NET Framework 4.0.

For a rich development environment, the following are recommended:

- Microsoft Visual Studio® development system 2010 SP1 (any edition).
- Silverlight 4.0 Tools for Visual Studio® 2010.

To run the Enterprise Library unit tests, you will also need:

- Visual Studio 2010 Premium, Visual Studio 2010 Professional, or Visual Studio 2010 Ultimate edition.
- Moq v4.0 assemblies.

To use the Stock Trader reference implementation, you will also need the following items installed:

- Microsoft SQL Server 2008 Express.

Installation Using NuGet and Extension Manager

Enterprise Library is now available via NuGet to install and update references to libraries in your projects. All Enterprise Library 5.0 components have been added as NuGet packages. The diagram below illustrates the Enterprise Library NuGet Package dependencies.

> **Note:** *NuGet is a free, open source developer-focused package management system for the .NET platform intent on simplifying the process of incorporating third party libraries into a .NET application during development. It helps you to quickly add references to .NET libraries such as Enterprise Library.*

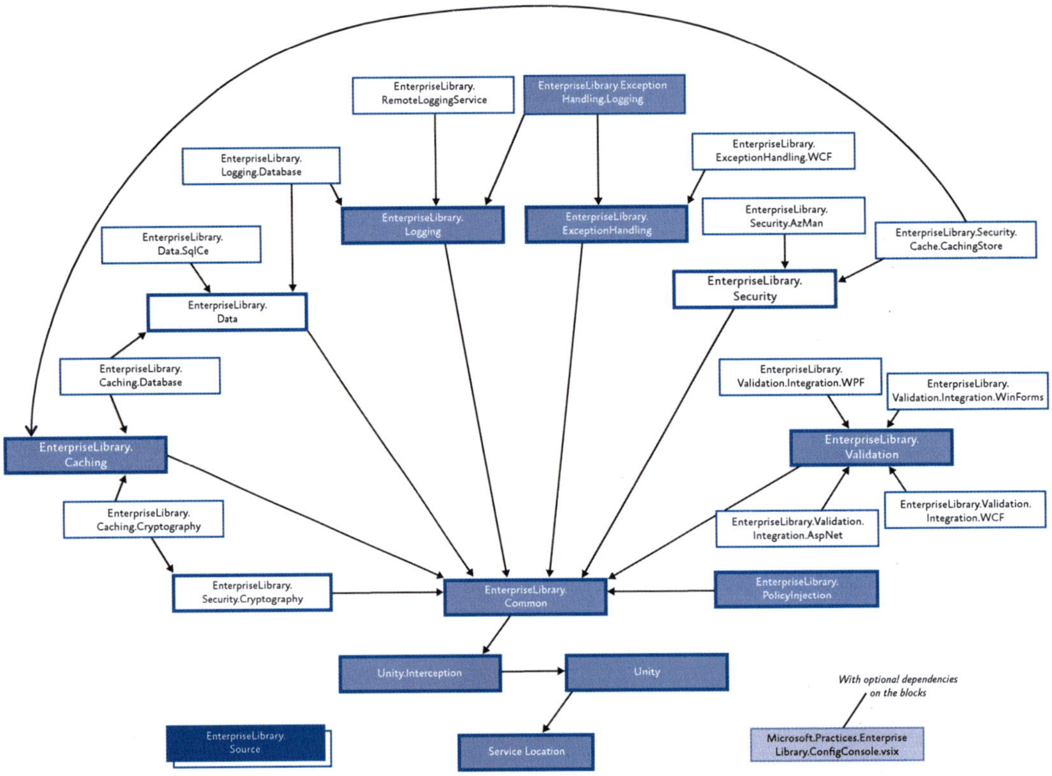

NuGet provides a great experience for adding references to Enterprise Library binaries to your project. For example, if you wish to apply the Enterprise Library Caching block, you can simply select and install the caching block from the NuGet Library Manager. It will automatically download and install the component and all of its dependencies, such as the **Microsoft.Practices.EnterpriseLibrary.Common** and the **Microsoft.Practices.Unity** assemblies.

For the assemblies that have both a Silverlight and a .NET Framework version, the packages will contain both versions (denoted in blue in the diagram). However, NuGet takes into account the type of project you are trying to add a reference to. If you have a Silverlight project selected, it will only add a reference to the Silverlight project, but if you have a .NET Framework project selected, then it will only add the reference to the .NET Framework project.

For example, when you download the Enterprise Library 5.0 Validation Application Block, it will download both the Silverlight and .NET Framework assemblies. If you try to add that reference to a Silverlight project, then only the reference to the Silverlight project will be added. For the Data Access Application Block, which is only available for the .NET Framework, the package will only contain the .NET Framework assembly. If you try to add the Data Access Application Block to a Silverlight project, you will get an error message.

> The NuGet packages do not contain the configuration console or any of the other tools. If you wish to use the configuration console, you can use the Extension Manager to download it. Any other tool, such as the MSMQDistributor, mergeconfiguration.exe, or installutilities.bat can only be obtained by installing the complete MSI.

Installing NuGet into Visual Studio

To install NuGet into Visual Studio, perform the following steps:

1. Start Visual Studio 2010.
2. Go to **Tools -> Extension Manager**.
3. From the left menu, select **Online Gallery**.
4. In the search box, type **NuGet** and wait for the list to filter.

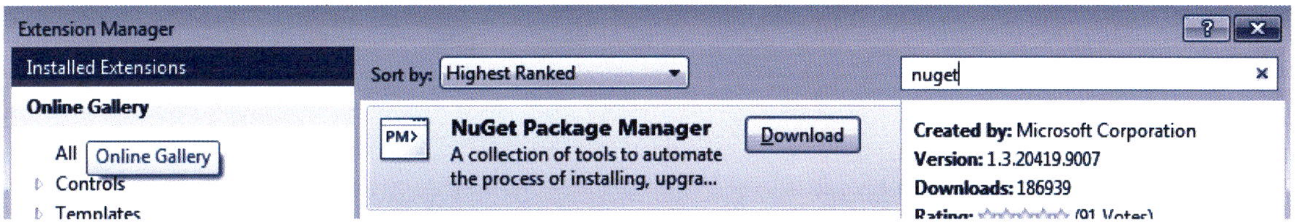

5. Select **Download** on the NuGet Package Manager.
6. Read the license agreement and if you agree, click **Install**.
7. Restart Visual Studio.

Adding Enterprise Library Components to Your Application

To add an Enterprise Library component to your application, perform the following steps:

1. Right click your project and select **Add Library Package Reference**.
2. From the left menu, select **Online**. If you are using a local repository, point to that feed.
3. In the search box, type **Enterprise Library** and wait for the list to filter.

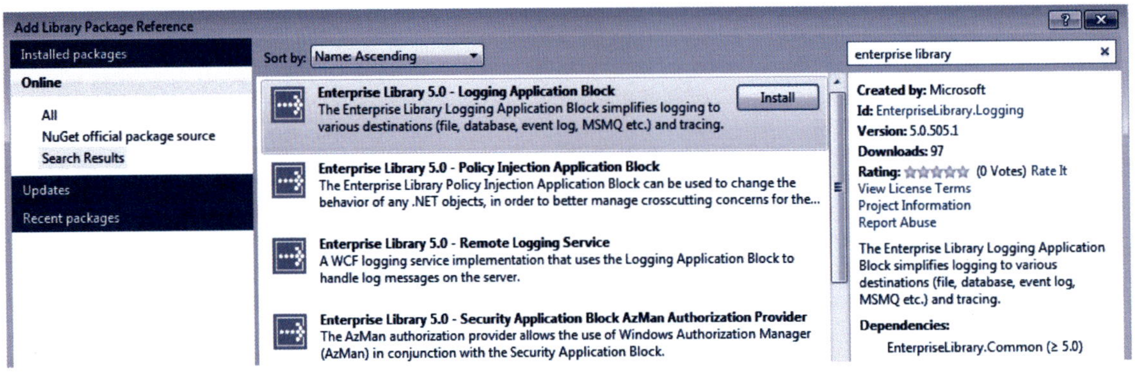

4. Select the block you wish to add and click **Install**.

NuGet will now download the required dependencies and store them in your solution folder under the folder named packages. When you install a NuGet package, it will also register the location of the binaries as a solution property, which is needed by the configuration console.

Database Scripts

There are two NuGet packages for Enterprise Library 5.0 Optional Update 1, that also contain database scripts. These are EnterpriseLibrary.Caching.Database and EnterpriseLibrary.Logging.Database. Once you have installed these packages, you can find the files in the following locations, relative to your solution file:

- /Packages/EnterpriseLibrary.Caching.Database 5.0.505.0/Scripts/CreateCachingDatabase.sql
- /Packages/EnterpriseLibrary.Logging.Database 5.0.505.0/Scripts/CreateLoggingDatabase.sql

Installing the Configuration Console

The configuration console can be installed from the extension manager:

1. Start Visual Studio 2010.
2. Go to **Tools -> Extension Manager**.
3. From the left menu, select **Online Gallery**.
4. In the search box, type **Enterprise Library** and wait for the list to filter.

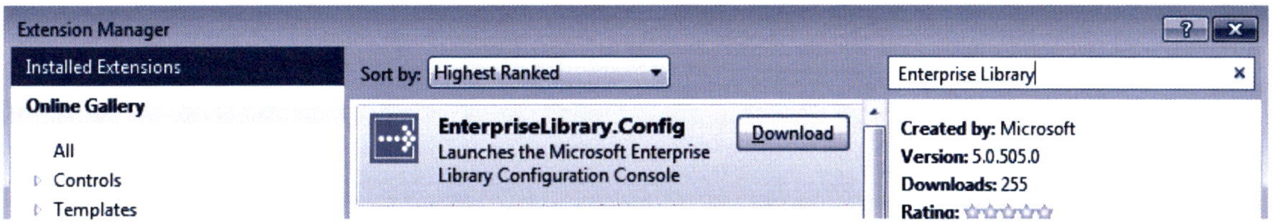

5. Select **Download** on the EnterpriseLibrary.Config item.
6. Read the license agreement and if you agree, Click **Install**.
7. Restart Visual Studio.

Alternatively, you can download the Visual Studio Extension package (.vsix) and install it.

The configuration console is now available when you right click a .config file. You can choose **Edit Server Configuration**, which will start the configuration console in the desktop profile. You can also choose **Edit Silverlight configuration file** on .config files other than Web.config, which will start the configuration console in the Silverlight profile.

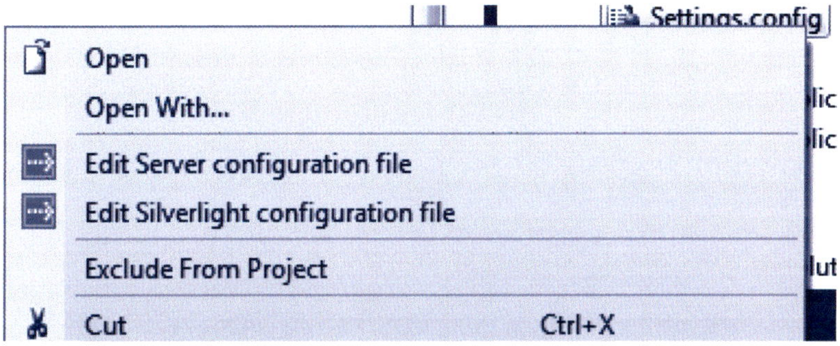

MSI Installation

If you don't wish to use NuGet install Enterprise Library, then you can also install Enterprise Library 5.0 Optional Update 1 and the Enterprise Library Silverlight Integration Pack using an MSI.

The installer for Enterprise Library 5.0 Optional Update 1 contains the updated versions of Enterprise Library 5.0 assemblies and the updated configuration console.

The installer for the Enterprise Library 5.0 Silverlight Integration Pack contains the binaries for Silverlight and the design time components for the configuration console.

Performing a Clean Installation

If you do not have Enterprise Library installed, then you can perform a clean installation. To do this, you will have to install the following MSI's:

- Enterprise Library 5.0 Optional Update 1.msi
- Enterprise Library 5.0 Silverlight Integration Pack.msi

Upgrading from Enterprise Library 5.0

If you have Enterprise Library 5.0 already installed, then you have two options for continuing:

1. Uninstall Enterprise Library 5.0 and then execute a clean installation.
2. Upgrade Enterprise Library 5.0 to Optional Update 1 from the command prompt, then install the **Enterprise Library 5.0 Silverlight Integration Pack.msi**

To upgrade Enterprise Library 5.0 to Optional Update 1 from the command prompt, you can perform the following steps:

1. Rename the **Enterprise Library 5.0 Optional Update 1.msi** file to **Enterprise Library 5.0.msi.**

Execute this command from the command prompt:

```
msiexec /i "Enterprise Library 5.0.msi" REINSTALLMODE=vomus
REINSTALL=ALL
```

This command will update your existing installation of Enterprise Library to Optional Update 1.

Upgrading from a Version Earlier than Enterprise Library 5.0

Enterprise Library 5.0 introduced many improvements over previous versions of Enterprise Library. It did however, introduce a couple of breaking changes. If you are currently using a version of Enterprise Library older than 5.0, we recommend that you use the **Migration Guide** to help you to upgrade to the latest version of Enterprise Library 5.0.

Installing the Learning Materials

The Enterprise Library 5.0 Silverlight Integration Pack ships with several learning materials:

- Documentation
- Stock Trader V2 Reference Implementation
- Quick Starts

Documentation

This guide and the API Reference are available for download from: http://go.microsoft.com/fwlink/?LinkID=217989

Installing the Stock Trader V2 Reference Implementation

The Stock Trader V2 reference implementation is available as a **separate download**.

To run the Stock Trader V2, perform the following steps:

1. Ensure you have the required pre-requisites installed.
2. Download the zip archive from CodePlex.
3. Extract the source code.
4. Open the StockTraderRI.sln using Visual Studio 2010.
5. Ensure the project StockTraderRI.Web is set as your startup project.
6. Press F5 to compile and run the Stock Trader V2.

QuickStarts

The Enterprise Library 5.0 Silverlight Integration Pack ships with the following QuickStart:

- WCF Data Services QuickStart

Source Code

The Source Code for Enterprise Library is available from **NuGet** or **CodePlex**. You can download the source code either as a zip file, or use one of the many **supported** source control clients to retrieve the source code.

The source code can be used to learn how Enterprise Library works, to use as a basis for custom extensions, or to create a custom signed version for Enterprise Library.

2 Stock Trader V2 Reference Implementation

The Enterprise Library 5.0 Silverlight Integration Pack ships with an example application, called Stock Trader V2, that demonstrates how you can apply Enterprise Library to a Microsoft® Silverlight® application. This application addresses some real-world challenges that developers face every day when building complex applications.

The Stock Trader V2 reference implementation may look familiar to you. We have based it on the Stock Trader reference implementation from another patterns & practices project, Prism.

> Prism provides guidance designed to help you more easily design and build rich, flexible, and easy-to-maintain Microsoft Windows® Presentation Foundation (WPF) desktop applications and Silverlight rich Internet applications (RIAs) and Windows Phone 7™ applications. Using design patterns that embody important architectural design principles, such as separation of concerns and loose coupling, Prism helps you design and build applications using loosely coupled components that can evolve independently but which can be easily and seamlessly integrated into the overall application. Such applications are often referred to as composite applications.

Limitations

The goal of Stock Trader V2 is to show techniques for applying Enterprise Library in various scenarios.

Stock Trader V2 is not meant to be a fully working implementation of a Stock Trading application. For example, while the Stock Trader uses web services and a database for certain situations, it also uses randomly generated data for other scenarios. You should view the Stock Trader V2 as a demonstration of how you can apply Enterprise Library to solve a number of common development challenges.

The Scenario

The company, Woodgrove Bank, has always been a big player in the stock trading software market. Over the last decade, they have grown to become a market leader with their stock trading application Stock Trader, affectionately called "The Old Stock Trader." Several thousand stock traders use this application daily, and together they trade more than $10 million in stocks every few hours.

Problems with the Old Stock Trader

While the old Stock Trader is still quite competitive with regards to its functionality, it's definitely showing its age:

- It was built as an ASP.NET 1.0 web application and didn't use any modern techniques for providing a great user experience such as AJAX, jQuery or Silverlight, and users were complaining about its usability and performance.
- It was not built to be very maintainable. One issue was the tight coupling of its components. Adding new features or changing existing logic was always challenging and often introduced more problems.
- It didn't provide much helpful diagnostic information. Diagnostic logging was only added here and there as an afterthought. Typically the only way problems with the application were detected was when angry users would notify the support team.
- The parts of the Old Stock Trader that allowed a user to edit his account holder information and to place orders had many problems. The validation logic was spread throughout the application and was riddled with duplicate code.

Because of these problems, about two years ago efforts to improve the old Stock Trader began. The following areas were improved:

- Enterprise Library was applied to help with exception handling and diagnostic logging and tracing. Exceptions were written to the Windows Event Log. There was not enough time to implement this everywhere, but at least the most critical areas would now report problems in a structured way.
- A couple of years ago, Woodgrove Bank began to use System Center Operations Manager (SCOM) to monitor the health of its applications. Now that the old Stock Trader also produced diagnostic information, SCOM could also be used to monitor its health.
- The logic that allowed a user to edit his account holder information was completely rewritten to use the `ADO.NET Entity Framework` and the Validation Application Block. There was not enough money in the budget to rewrite the order processing functionality as well, so this remained largely unchanged.

Stock Trader V2 Overview

While the improvements did much to increase the stability, manageability, and performance of the Stock Trader, the user experience still didn't meet the end users' expectations. For that reason, last year the management of Woodgrove Bank decided to build a new version of the Stock Trader, using Silverlight as a front end. It was very important to Woodgrove Bank's management that the investments that have been made to improve the old Stock Trader would not be wasted, therefore diagnostic logging and tracing for the new Stock Trader should build upon the foundation that was laid by the old Stock Trader, and the data model and validation rules for the account holder module would have to be reused as much as possible.

Stock Trader V2 also needed a new and fresh appearance and feel. This is what the new stock trader looks like:

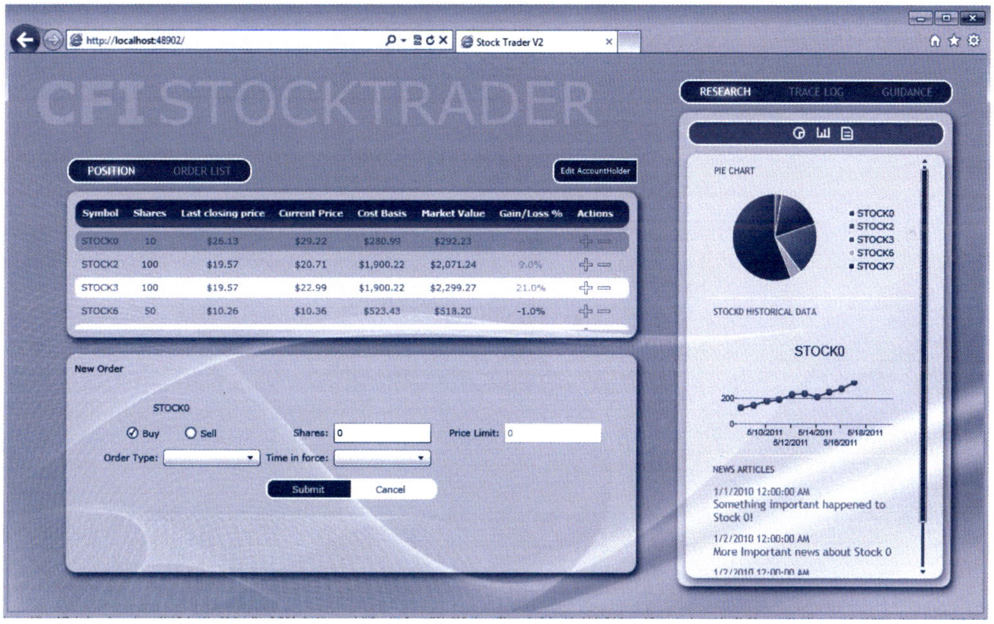

In this section, we'll provide an overview of the new Stock Trader V2 application.

Note: *Each module has a theme that maps to a particular aspect of Enterprise Library. For example, the position module focuses on caching and the Account Holder Info Module focuses on validation in combination with Windows Communication Foundation (WCF) Data Services. The exception to this is Exception Handling and Logging, which are implemented everywhere in the application.*

Overview of Your Account Position

When the Stock Trader reference implementation starts, users will first see the account position summary screen. Here they can get an overview of all the stocks they currently own, and see their cost basis and current market value.

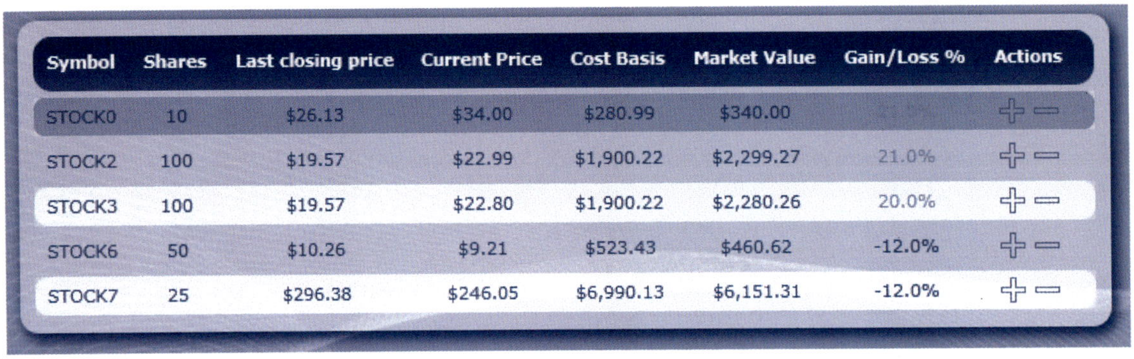

Symbol	Shares	Last closing price	Current Price	Cost Basis	Market Value	Gain/Loss %	Actions
STOCK0	10	$26.13	$34.00	$280.99	$340.00	[illegible]	+ −
STOCK2	100	$19.57	$22.99	$1,900.22	$2,299.27	21.0%	+ −
STOCK3	100	$19.57	$22.80	$1,900.22	$2,280.26	20.0%	+ −
STOCK6	50	$10.26	$9.21	$523.43	$460.62	-12.0%	+ −
STOCK7	25	$296.38	$246.05	$6,990.13	$6,151.31	-12.0%	+ −

To improve startup performance, the account position screen will cache the current account position to isolated storage. Therefore, the data is also available, even if you restart the application. The account position will expire at the end of the day. To refresh the data, you can select Reload from the context menu of the current account position view.

Selecting a particular stock will trigger the market history trend line and news module to be displayed. As soon as the current account position is loaded, the Stock Trader will automatically start proactively caching the market history.

Note: *The main goal of the Account Position Summary is to demonstrate:*

- *Caching to isolated storage*
- *Exception handling with WCF Data Services*
- *Logging*
- *Tracing with asynchronous messages*

Analyze Market Trends

When a user selects a particular stock in the Account Position module, the trend line view will show the history of that particular stock in a graph.

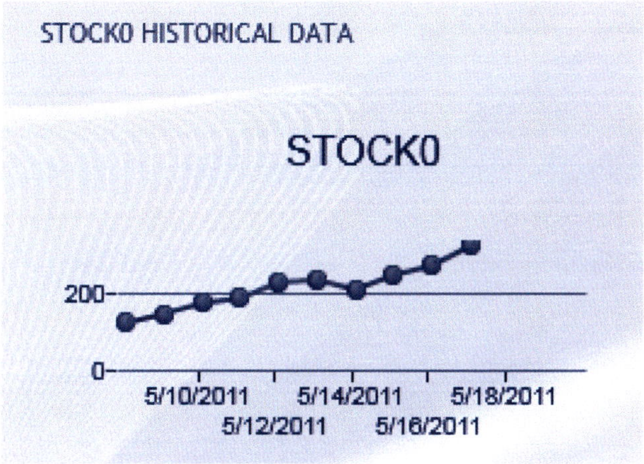

To improve the startup performance of the trend line view, the information is proactively cached to isolated storage. That means that the cache is populated on a background tread. Since the data is cached to isolated storage, it will also be available when the application starts.

> **Note:** *The main goal of the market history is to demonstrate:*
> - *Proactive caching*
> - *Caching to isolated storage*
> - *Logging*

Read Relevant News Articles

When a user selects a particular stock, a list of relevant news articles is shown for that stock. Selecting the article will cause its full text to be displayed.

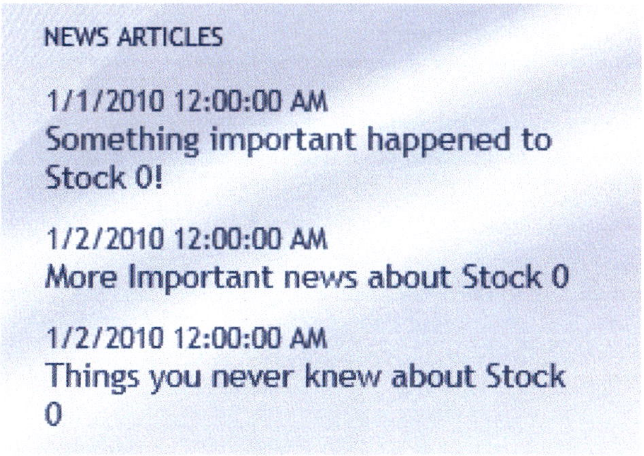

> **Note:** *The main goal of the news module is to demonstrate:*
> *- Caching to the in-memory cache*
> *- Using Unity Interception to asynchronous services.*
> *- Using Unity Interception to apply exception handling to asynchronous services.*

Place Orders to Buy Or Sell Stock

From the list of stocks in the account position view, users can choose to place orders to buy or sell. An order can consist of multiple requests to buy or sell a particular stock.

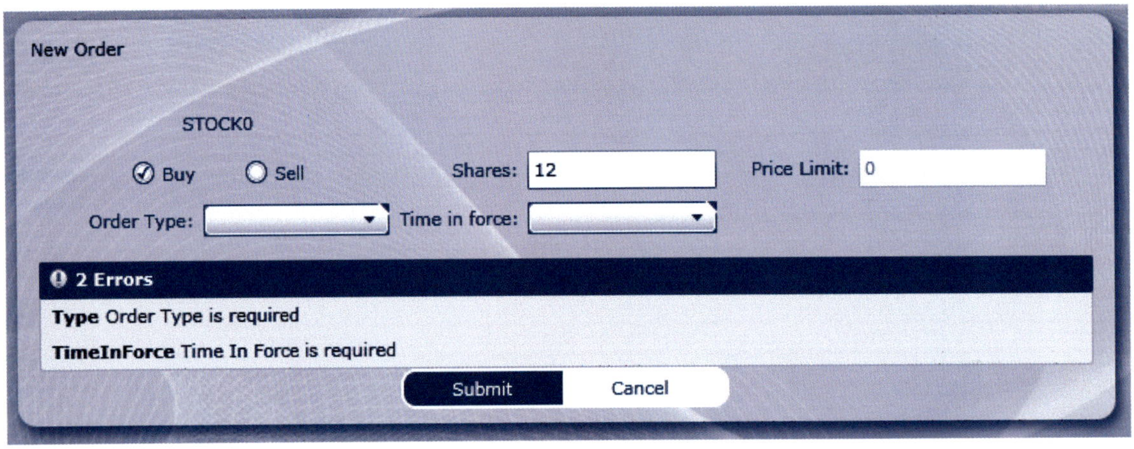

Validation of input and business rules is an important part of the user experience. While the user is editing, he should get visual feedback of any errors, but this feedback should not interrupt him in his work. If there are still errors when the user tries to submit, he should get a message saying that there were additional validation errors.

The user should not be able to submit an order that has errors. Some validation rules should be implemented as warnings and the user should be able to continue submitting after a warning.

The business group also wanted the ability to add additional validation rules without having to recompile and redeploy the application.

> **Note:** *The goal of the orders module is to demonstrate:*
> *- The defining of validation rules in code*
> *- The defining of validation rules in configuration*
> *- The use of validation rule sets*
> *- Validation in a WCF Data Services application*
> *- Complex type validation*

- Building a ViewModel using a base class
- Logging
- Asynchronous tracing

Editing Current Account Holder Information

Users should be able to view and update their own account information, including their personal information, preferred risk tolerance, and their bank accounts.

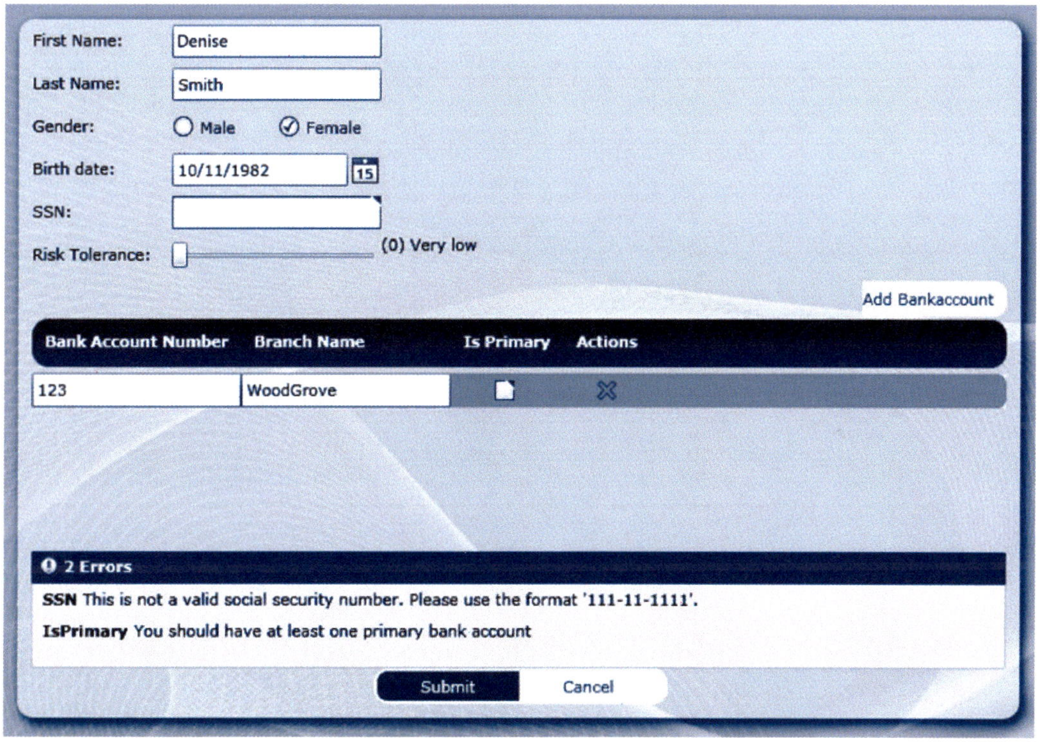

Input and business rule validation is important here too. And while users can add multiple bank accounts, only a single bank account can be marked as primary.

Note: *The goal for the account holder module is:*
- Validation using Enterprise Library
- Sharing validation rules between client and server
- Validation in combination with WCF Data Services (Technique demonstrated also works with regular WCF services)
- Building view models

Guidance Module

The guidance module has been added to provide a description of the functionality for the stock trader from within the application. It helps explain the workings of the application.

Capture Diagnostic Information

One of the reasons the old Stock Trader was difficult to maintain was the lack of good diagnostic information. For Stock Trader V2, there were a couple of requirements for diagnostic information:

- Developers should be able to view a developer trace log. This log should display all diagnostic information that is gathered by the application. Developers should be able to group this information by activity or category.

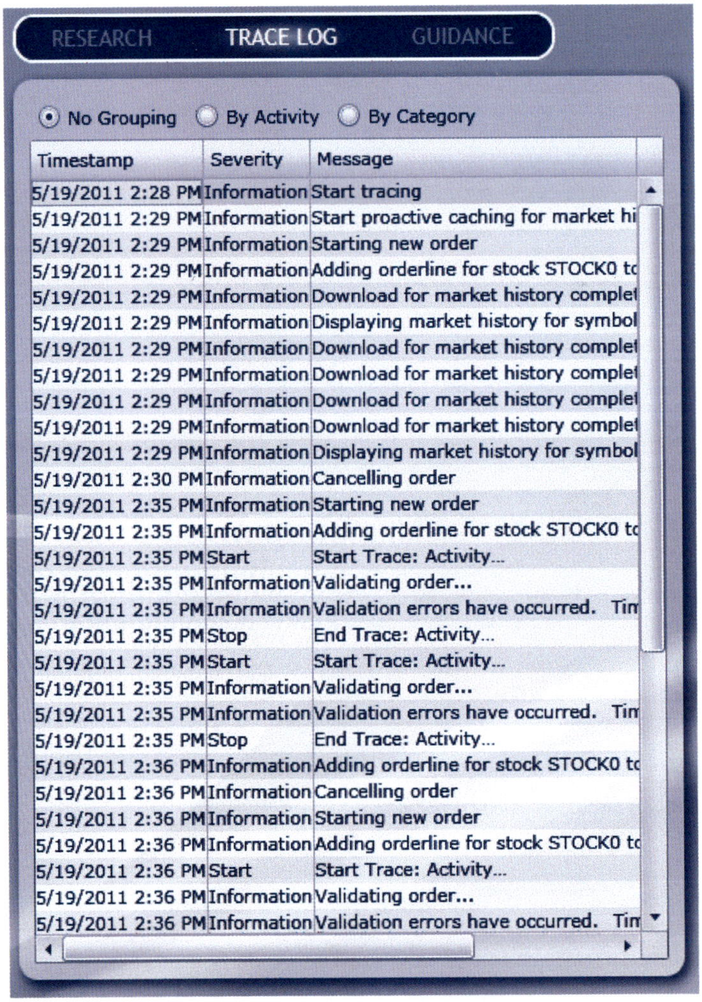

- Power users should be able to see a log file with important log messages. This log file should be stored on the end user's hard drive so that it is not lost when the application restarts. Power users can reach the log view by adding #/IsolatedStorageLog to the URL, so that the URL looks like this: http://localhost:48902/Default.aspx#/IsolatedStorageLog.

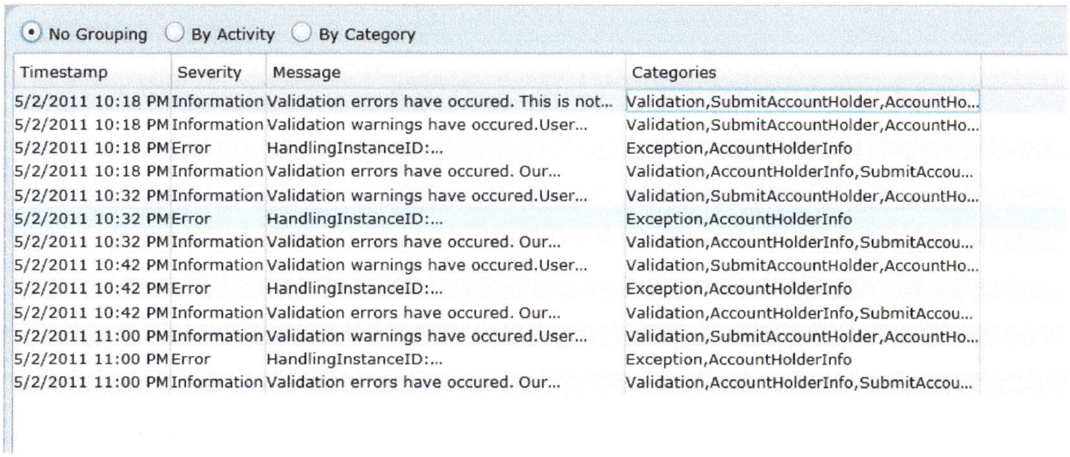

No Grouping / By Activity / By Category

Timestamp	Severity	Message	Categories
5/2/2011 10:18 PM	Information	Validation errors have occured. This is not...	Validation,SubmitAccountHolder,AccountHo...
5/2/2011 10:18 PM	Information	Validation warnings have occured.User...	Validation,SubmitAccountHolder,AccountHo...
5/2/2011 10:18 PM	Error	HandlingInstanceID:...	Exception,AccountHolderInfo
5/2/2011 10:18 PM	Information	Validation errors have occured. Our...	Validation,AccountHolderInfo,SubmitAccou...
5/2/2011 10:32 PM	Information	Validation warnings have occured.User...	Validation,SubmitAccountHolder,AccountHo...
5/2/2011 10:32 PM	Error	HandlingInstanceID:...	Exception,AccountHolderInfo
5/2/2011 10:32 PM	Information	Validation errors have occured. Our...	Validation,AccountHolderInfo,SubmitAccou...
5/2/2011 10:42 PM	Information	Validation warnings have occured.User...	Validation,SubmitAccountHolder,AccountHo...
5/2/2011 10:42 PM	Error	HandlingInstanceID:...	Exception,AccountHolderInfo
5/2/2011 10:42 PM	Information	Validation errors have occured. Our...	Validation,AccountHolderInfo,SubmitAccou...
5/2/2011 11:00 PM	Information	Validation warnings have occured.User...	Validation,SubmitAccountHolder,AccountHo...
5/2/2011 11:00 PM	Error	HandlingInstanceID:...	Exception,AccountHolderInfo
5/2/2011 11:00 PM	Information	Validation errors have occured. Our...	Validation,AccountHolderInfo,SubmitAccou...

- The IT support organization needs to be able to detect problems with the stock trader. When an unknown problem occurs on the Silverlight client, the application must report the problems to the server where they need to be added to the server's Windows Event Log. System Center Operations Manager will monitor the server's event logs and report any problems to the end users.

Stock Trader V2 Architecture

In this topic, we'll discuss the overall architecture of the Stock Trader V2. An LOB Silverlight application, like the StockTrader V2 reference implementation, typically consists of a Silverlight client and a web application where the Silverlight application is hosted.

The following diagram shows the general layering that's applied to Stock Trader V2.

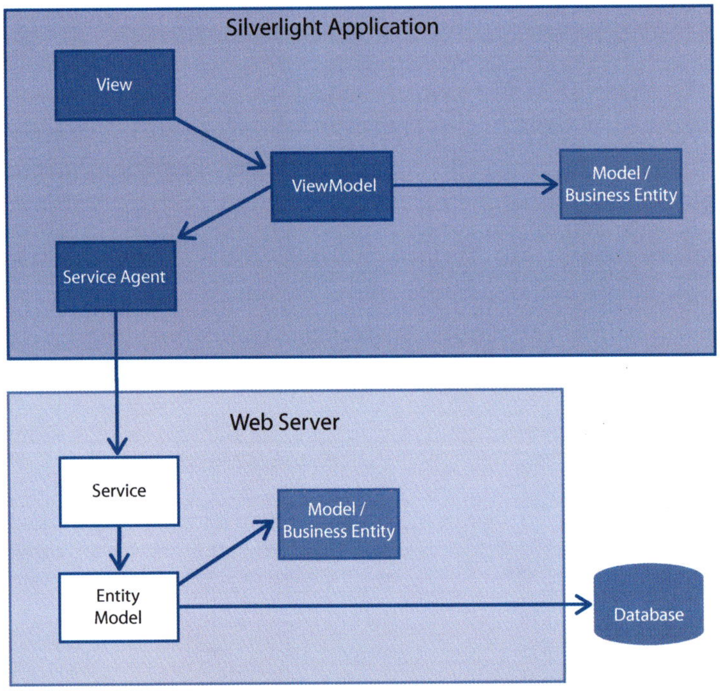

The user interface follows the Model View ViewModel pattern. The goal of this pattern is to separate the view from the view's logic and data. The reason for the separation is that the view is typically a user control and difficult to test using automated unit tests. The logic is encapsulated in a special ViewModel class, which is then easily unit tested.

You can read more about the Model View ViewModel pattern in the **MVVM documentation** from Prism on MSDN, and in **Karl Shifflet's In the Box Training**.

The Silverlight application accesses several web services, which are exposed by the web server. These web services are implemented using various technologies, such as WCF, WCF RIA Services and WCF Data Services. The logic to interact with these services has been encapsulated in special service agent classes. This isolates the View Model from web service-specific implementation details and also improves the testability of the View Model.

The business entities are shared between the client and the server. Each web service technology has its own method for achieving this. For example, WCF RIA Services uses a code generator to generate the business entities on the client.

On the server, the web services use an entity model to define the business entities and the object relational mapping to the database. The business entities also typically define certain validation rules.

Modular Development

The Silverlight part of the Stock Trader has been developed as a modular application. That means that it has been divided into several discrete modules. Each module addresses a functional area, such as news, account position, or orders.

> Dividing the Stock Trader V2 into several discrete modules turned out to be a really good idea. It allowed each module to be developed, tested, and deployed in isolation. It also helped us to implement a clear separation of concerns between the application's horizontal capabilities, such as logging or exception handling, and the vertical capabilities, such as processing an order or editing an account holder.

The following diagram shows the modules for the Stock Trader V2 reference implementation:

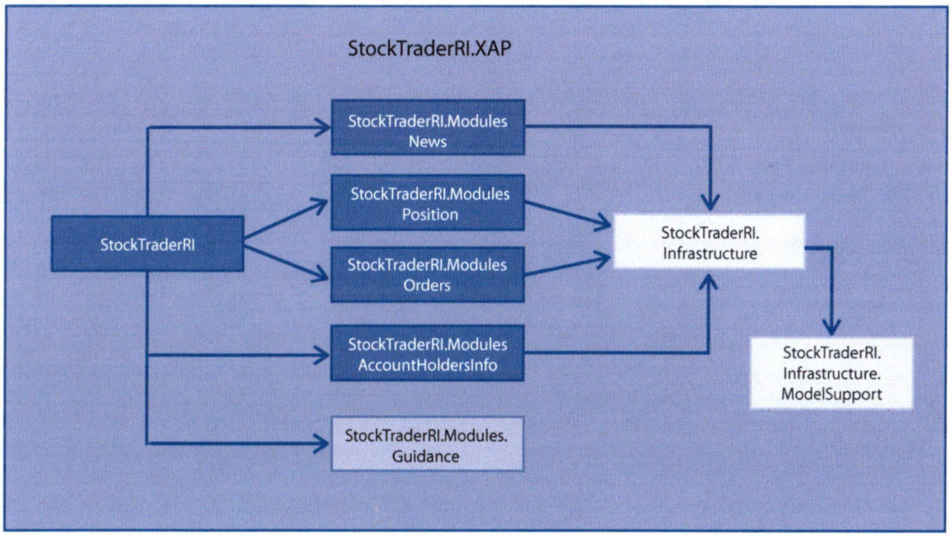

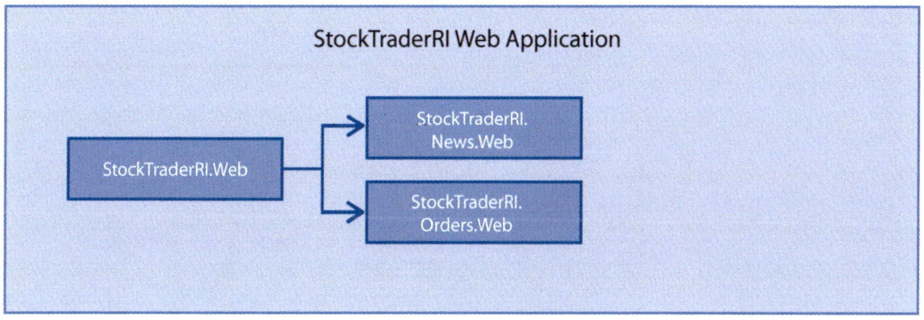

The following assemblies will be packaged into the StockTrader RI.XAP:

- **StockTraderRI.** This is the main shell, which references all modules and knits them together into a single application. It also contains the views and styles that determine the general look and feel for the application. The shell knits together these functional modules:
    - StockTraderRI.Modules.News
    - StockTraderRI.Modules.Position
    - StockTraderRI.Modules.Orders
    - StockTraderRI.Modules.AccountHolderInfo
- **StockTraderRI.Modules.Guidance.** The guidance module is used to describe the functionality of the Stock Trader V2 reference implementation from within the reference implementation itself.
- **StockTraderRI.Infrastructure.** The infrastructure assembly contains generic functionality that is used by all modules.
- **StockTraderRI.Infrastructure.ModelSupport.** The Model Support assembly contains classes that help you with the most common tasks when using the Model-View-ViewModel pattern, such as implementing **INotifyDataErrorInfo** and **INotify PropertyChanged** on your models.

The web application consists of the following parts:

- **StockTraderRI.Web.**
  The StockTraderRI.Web is a web application that hosts all the web services. It also hosts the Silverlight XAP file and the main page that is used to launch the XAP file. The WCF RIA Services for news and orders have been placed into separate projects:
    - StockTraderRI.Orders.Web
    - StockTraderRI.News.Web

3 Configuration

One of the strengths of Enterprise Library has always been the flexibility it provides through its configuration settings. By making changes to an application's configuration you can, for example, change the way the Enterprise Library logs exceptions or handles validation.

In the Enterprise Library 5.0 Silverlight® Integration Pack, the process for configuring Enterprise Library is largely the same as in Enterprise Library 5.0. You can either create a configuration file using the configuration console, create the configuration files by hand, or programmatically configure Enterprise Library.

In this chapter, we'll describe several options for creating configuration files with the Enterprise Library 5.0 Silverlight Integration Pack, for example by using the configuration console. We'll also discuss several things to keep in mind when creating configuration files. We'll also describe several options for consuming the configuration files in a Silverlight application. At the end of this chapter, we'll have a look at how the Stock Trader V2 Reference Implementation uses configuration.

Configuring Enterprise Library

The most common way to configure Enterprise Library is to use the configuration console. This tool provides an easy-to-use graphical user interface for creating configuration files. The following image shows an example of a configuration file as it is shown in the configuration console.

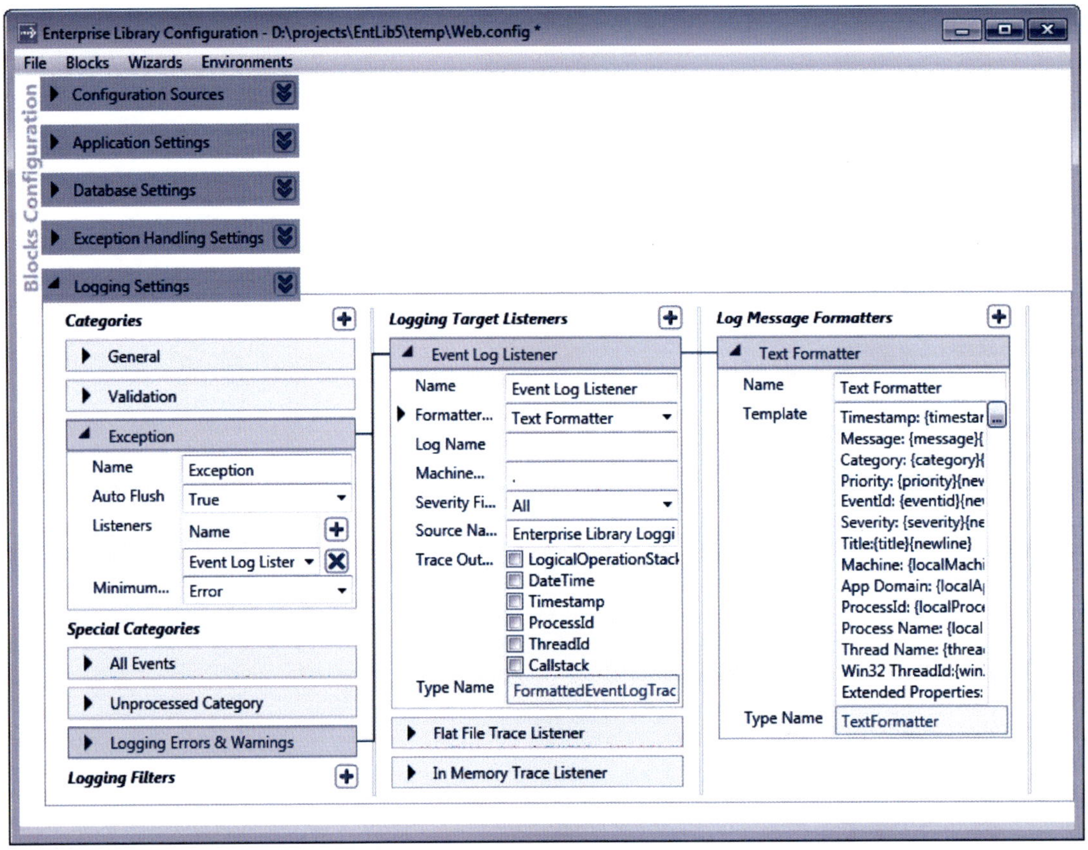

This configuration file shows how exceptions and logging errors and warnings are being logged to the Windows Event Log. The Text Formatter is used to create an easy-to-read log message. There is a section in the configuration console for each application block.

For more information on creating application configuration files using the configuration console, please refer to the chapter on **Using Configuration Tools** in the Enterprise Library documentation on MSDN.

Configuration in XAML Files

The Enterprise Library 5.0 Silverlight Integration Pack does not rely on application configuration files, because the System.Configuration namespace is not available in Silverlight. Instead, it stores configuration information in XAML configuration files. The following snippet shows an example of such a XAML configuration file, which configures the Logging Application Block to write log messages to isolated storage.

```
XML
<el:ConfigurationDictionary
  xmlns="http://schemas.microsoft.com/winfx/2006/xaml/
  presentation"
  xmlns:x="http://schemas.microsoft.com/winfx/2006/xaml"
  xmlns:el="http://schemas.microsoft.com/practices/2011/
  entlib">
  <el:LoggingSettings DefaultCategory="default"
    x:Key="loggingConfiguration"
    LogWarningWhenNoCategoriesMatch="False">
    <el:LoggingSettings.TraceListeners>
      <el:IsolatedStorageTraceListenerData
        Name="isolated"
        RepositoryName="LogMessagesInIsolatedStorage"
        MaxSizeInKilobytes="256"/>
    </el:LoggingSettings.TraceListeners>
    <el:LoggingSettings.TraceSources>
      <el:TraceSourceData Name="default"
        DefaultLevel="Warning">
        <el:TraceSourceData.TraceListeners>
          <el:TraceListenerReferenceData
            Name="isolated"/>
          </el:TraceSourceData.TraceListeners>
    </el:LoggingSettings.TraceSources>
  </el:LoggingSettings>
</el:ConfigurationDictionary>
```

One of the drawbacks of using XAML for storing configuration is that it's more verbose than the traditional XML configuration files. Fortunately, you don't have to edit the XAML files by hand. In the next chapter, we'll describe how you can generate the XAML configuration files from application configuration files, which can be edited using the configuration console.

Options for Generating XAML Configuration Files

While it's possible to edit the XAML configuration files used by the Enterprise Library 5.0 Silverlight Integration Pack by hand, it's much easier to generate them from application configuration files, which can be edited by the Enterprise Library Configuration Console.

The following diagram shows several ways for generating XAML configuration files:

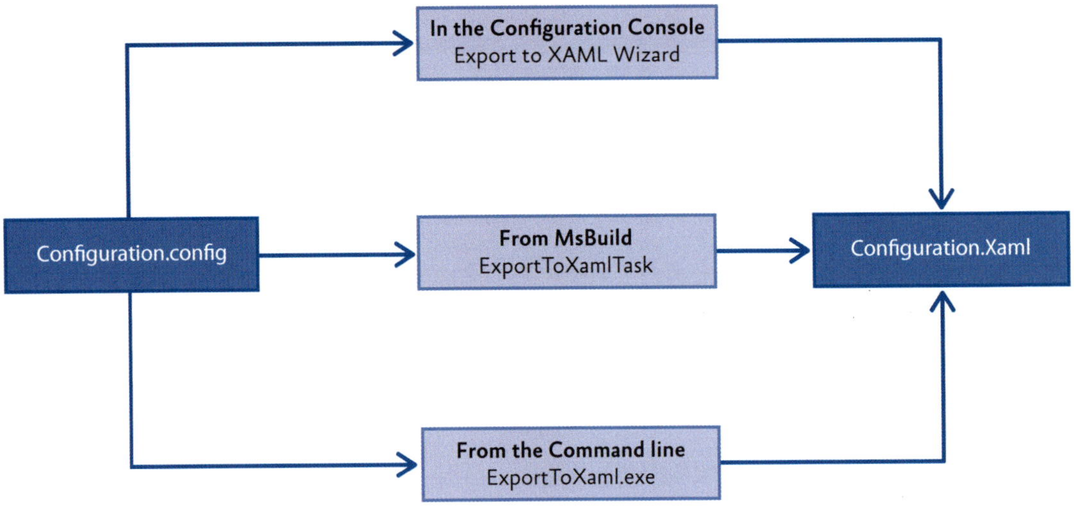

When you have created an application configuration file, you can convert it to a Silverlight XAML configuration file using a wizard in the configuration tool. There are also two options for automating this XAML generation process: using an MSBuild task, which allows you to integrate the XAML generator in an automated build process and using a command line executable, which makes it easy to execute the XAML generator from a script.

Creating Application Configuration Files Using the Configuration Console

The process for creating configuration files is largely the same as it was with Enterprise Library 5.0. First, you'll start up the configuration console, and then add configuration sections for the required application blocks, update the configuration, and save the file. For the Enterprise Library 5.0 Silverlight Integration Pack, this process is largely the same. However, depending on how you installed the configuration console, you will get a slightly different experience.

Configuration Console with Visual Studio Integration

If you have installed the configuration console from the Extension Manager, then you can launch the configuration console from Microsoft® Visual Studio®, by right clicking on an application configuration file. It also allows you to start the console with a new feature called profiles, which allows the console to only show application blocks that make sense for a particular environment. If you are creating a configuration file that will be consumed by the server, then you shouldn't be able to add application blocks that only work in Silverlight, and vice versa.

When your configuration file is in a web application project, then you can start the console with: **Edit Server configuration** (the desktop profile, which shows only the application blocks that work with the Microsoft .NET Framework) and **Edit Silverlight configuration** (the Silverlight profile, which only shows the application blocks for Silverlight). However, when your configuration file is in a Silverlight project, then you will only be able to select **Edit Silverlight Configuration** and start with the Silverlight profile.

You can edit an application configuration file from a Silverlight using this Visual Studio integration. However, the Silverlight project template will not allow you to add an application configuration file. If you want to add a configuration file to a Silverlight project so that you can use that file to generate a XAML configuration file, you will need to add the file as a regular XML (.xml) file, then rename it to an application configuration file (.config) file and add an empty <configuration/> element to it. Then you can edit it using the configuration console with Visual Studio integration.

Stand-alone Configuration Console

If you have installed the configuration console using the Microsoft Enterprise Library 5.0 installer, then you will not be able to start the configuration console from Visual Studio. You will also not be able to use the profiles, so you will see all the application blocks that are currently installed.

The stand-alone configuration console is compiled for different platforms: .NET Framework 4.0, .NET Framework 4.0 32-bit, .NET Framework 3.5. If you want to edit Silverlight configuration files, you will need to select the .NET Framework 4.0 32-bit version for the configuration console, because Silverlight 4 only supports 32-bit applications.

Selecting Types in the Configuration Console

In certain situations, the configuration console allows you to select certain types. For example, while configuring an exception handling policy for the Exception Handling Block, you will need to select the type of exception that you want to handle. There are some limitations to which types the configuration console can select.

The configuration console uses reflection to load the types from the assembly. To do this, the configuration console must actually load the assembly into the appdomain. Silverlight has support for binary compatibility, though it will not work on any type. You can only use types that reference only the following assemblies:

- Mscorlib
- System
- System.Core
- System.ComponentModel.Composition
- Microsoft.VisualBasic

Usually, this means that you can select Silverlight exception types, but custom extensions, such as a custom exception handler you can likely not select from the configuration console. The section on Using

Custom Enterprise Library Extensions describes how you can use these.

> **Note:** *When you select an exception formatter to export to XAML, you will need to take an additional step for this to be exported correctly. The runtime only expects a partial type name, but the XAML generation logic needs a fully qualified name. In order for the export to XAML logic to work, you will either need to change the type name to the fully qualified name, or reselect the type using the type picker.*

In Stock Trader V2, we're using this feature to share validation configuration between client and server. The validation configuration was stored in a separate configuration file, called validation.config. The web.config uses this file as a file-based configuration source. We're also using this file to generate the validation.xaml file. And we've built a mechanism in Stock Trader V2 to read and merge multiple configuration files.

Using Non-default Configuration Sources

The Enterprise Library configuration tool allows you to store configuration information in a different location than application configuration files. For example, you can store your configuration in an external XML file by using a file-based configuration source.

Only the wizard in the configuration console will honor the non-default configuration sources. The other tools, such as the MSBuild task, cannot use non-default configuration sources. That is why this option has been removed from the Silverlight profile. It is still present in the desktop profile though.

However, if you store the configuration for a particular block in a separate file using the file-based configuration source, then you can use the XAML generation tools directly on the separate configuration file.

Please refer to the online documentation on **Non-Default Configuration Sources** on MSDN.

Generating XAML Configuration Using the Configuration Console

The most direct option for creating XAML configuration files is using the configuration console. The configuration console contains a wizard that will take the information in the configuration file and export it to XAML.

It is possible for the configuration wizard to encounter problems during the generation process. For example, if you have configured a trace listener that is not supported in Silverlight, or have selected a type that does not exist in Silverlight, then the export wizard will report this as a warning. If you choose to continue, then the Export to XAML wizard will ignore the sections that cannot be converted to XAML. The application settings and database, which are added by default by the configuration console, will also be ignored by the wizard, since they are not supported in Silverlight.

The export wizard does not support exporting the configuration for custom blocks to XAML. This feature is supported by the command-line tool and the MSBuild task, so if you use custom configuration blocks, you should choose these options.

CREATING A XAML CONFIGURATION FILE USING THE WIZARD

1. Launch the Configuration Console.
2. Select **File a Open** to open a configuration file.
3. Select **Wizards a Export To XAML** to export the file to XAML.

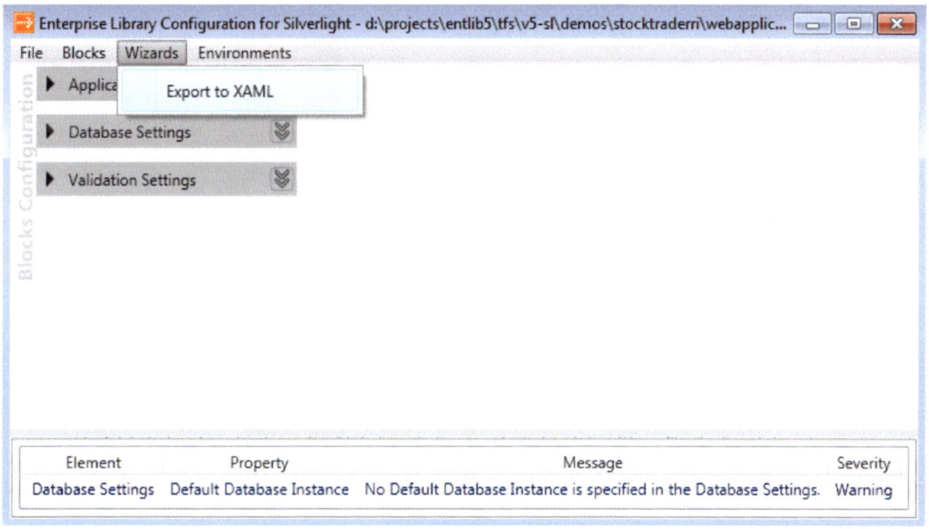

4. Enter the output filename in the wizard and press **Finish**. If you don't select a location and just enter a filename, then the file will be placed in your My Documents folder.

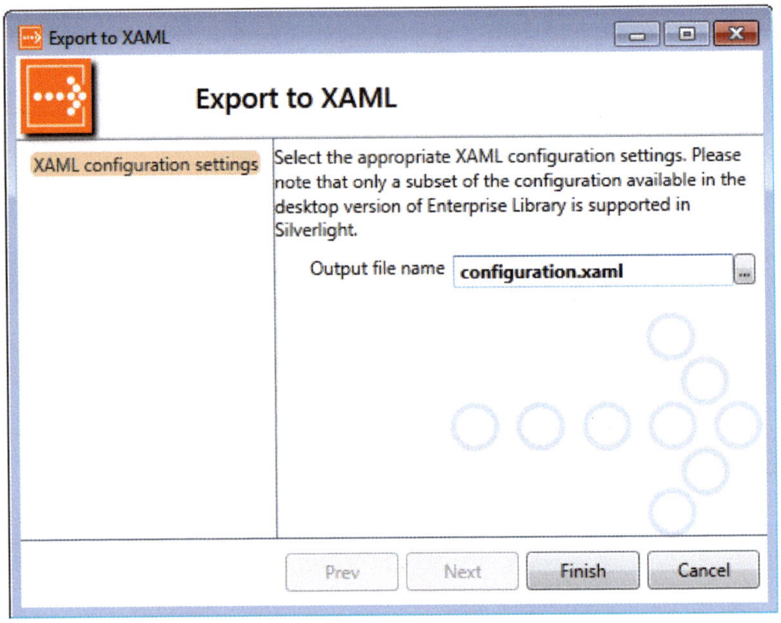

5. Include the configuration file in your Silverlight application, as described in the chapter on **Consuming XAML Configuration from a Silverlight application.**

Generating XAML Using the MSBuild Task

The Enterprise Library 5.0 Silverlight Integration Pack also ships with an MSBuild task that you can use to automate the generation of XAML configuration files in a build script.

The following shows an example of an MSBuild project file:

```
XML
<Project xmlns='http://schemas.microsoft.com/developer/msbuild/
   2003' DefaultTargets='Translate'>
   <UsingTask
      AssemblyFile='Microsoft.Practices.EnterpriseLibrary.
      Configuration.ExportToXaml.MsBuild.dll'
      TaskName='ExportToXamlTask'/>
   <Target Name='Translate'>
      <ExportToXamlTask Assemblies='$(PathToBinaries)'
         Source='Configuration.config' Destination=
         'Configuration.xaml'>
         <Output TaskParameter='EffectiveDestination'
            PropertyName='EffectiveDestination'/>
      </ExportToXamlTask>
```

```
    <Message Text='The value for output is
      $(EffectiveDestination)'/>
  </Target>
</Project>
```

This MSBuild file first configures the ExportToXamlTask, which resides in the assembly called Microsoft.Practices.EnterpriseLibrary. Configuration.ExportToXaml.MsBuild.dll. It then creates a new Target that will execute the ExportToXamlTask. This task supports the following parameters:

- **Assemblies (required).** You will need to specify where the binaries, required for the translation process, are located.
- **Source (required).** The name of the source application configuration file that should be converted to XAML.
- **Destination.** Allows you to specify the name and location for the generated XAML file.
- **EffectiveDestination (output).** Returns the actually used output file. This is useful if you don't specify a destination and want to use the output of this file further in the process.
- **OverwriteDestination.** Allows you to specify whether the output file should be overwritten.
- **Verbose.** Display all output, including information messages.
- **Sections.** Name of a section to include in the conversion, in addition to the Enterprise Library sections. Allows you to specify additional custom configuration sections that you want to have converted.
- **NameMappings.** Changes the name of the XAML element for the specified type (using the full type name, with no assembly name) to the specified expanded XAML name (using the "{namespace}element" syntax).
- **TypeMappings.** Indicates a type name translation for configuration settings containing type names.

Generating XAML Configuration Files Using the Command Line

It is also possible to generate XAML configuration files from the command line. This can be useful for automating the configuration generation if you're not using MSBuild.

The syntax for the command-line tool is:

```
ExportToXaml.exe "source file" [options]
```

There are several options you can specify:

Option	Description
-out:destination	Name of the output XAML file. Defaults to the name of the source file with a .xaml extension.
-force	Forces the file to be overwritten, even if the file is read-only.
-verbose	Display all outputs, including information messages.
-assemblies:path	Path to the folder with the assemblies required to load the configuration file.
-section:section	Name of a section to include in the conversion, in addition to the Enterprise Library sections. Allows you to specify additional custom configuration sections that you want to have converted.
-nameMapping:type=xname	Changes the name of the XAML element for the specified type (using the full type name, with no assembly name) to the specified expanded XAML name (using the "{namespace}element" syntax).
-typeMapping:type=mapping	Indicates a type name translation for configuration settings containing type names.
-translator:type=translator	Indicates a custom translator for a type. Requires the full name of the source type and the assembly-qualified name of the translator type.

Using Custom Enterprise Library Extensions

A common scenario when using Enterprise Library is to create and use your own custom extensions. For example, you might like to create a custom exception handler or a custom call handler. The Enterprise Library 5.0 Silverlight Integration Pack supports the usage of custom extensions, just like Enterprise Library 5.0.

However, as mentioned before, there are some limitations to which types you can select using the configuration console, due to the fact that the configuration console cannot load all Silverlight types. These limitations usually don't allow you to select a custom extension that is only present in Silverlight in the configuration console.

If you need the ability to select custom extensions from the configuration console, then you can perform the steps described in the following paragraph to create .NET Framework version of the custom extension. That version can then be loaded into the configuration console normally.

Place Custom Extensions in a .NET Assembly

If you want to add design-time support for your custom Silverlight extensions in the configuration console, you can do so by creating a desktop version of the custom types you'd like to configure. The following diagram shows an example of how this can work:

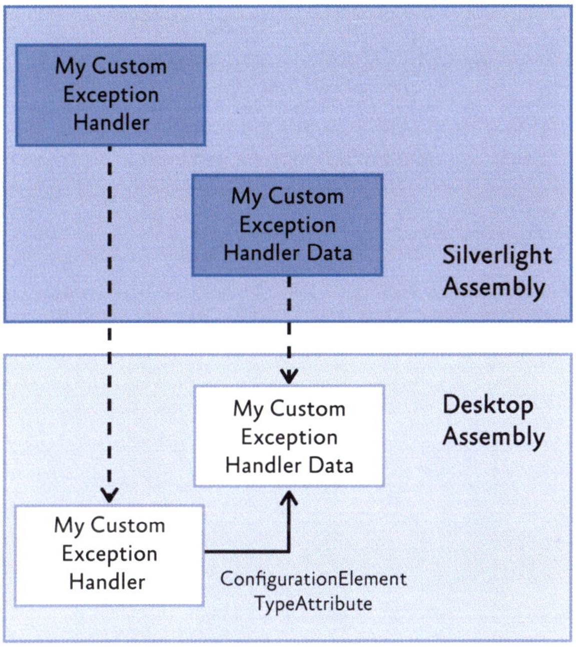

This diagram shows how you can create a custom exception handler, with its exception handler data, but would work the same for custom logging handlers.

To create a custom exception handler for the Enterprise Library 5.0 Silverlight Integration Pack, you'll need to create both a custom exception handler and its exception handler data. If you want to select the custom exception handler from the configuration console, you will also need to create a .NET Framework assembly that contains the same classes. Here are the guidelines to follow when creating a desktop version of your classes:

- The name of the type in the desktop assembly should match the name of the type in the Silverlight assembly.
- The desktop version of the custom exception handler should reference the custom exception handler data using the **Configuration Element Type Attribute**.
- If the namespace or assembly name differs between the desktop and Silverlight assembly, then it is recommended you use XAML

aliases. More information on XAML aliases can be found in the section on **Adding XAML Aliases.**

- If you don't want to use the XAML aliases, you can use the nameMappings property on the ExportToXAML MSBuild task and the ExportToXAML command-line tool. This property is not available in the wizard in the configuration console.
- The desktop version of the custom exception handler and custom exception handler data don't need to have actual implementations if you are just using them to allow for selection of the type in the configuration console. However, the easiest way to create the desktop version of the custom handler would be to create a Visual Studio linked file in the desktop project that links to the file in the Silverlight project. This ensures that the Silverlight and desktop versions will always have the same type names and namespaces.

Adding a XAML Alias

Having your Silverlight and desktop versions of the custom extensions in differently named assemblies or different namespaces can cause issues for the XAML configuration generation process The XAML generator, which is a desktop application, can easily read the desktop assemblies with their namespaces, but has no idea what the assembly name or namespace in Silverlight is going to be. To solve this, you can use XAML namespace aliases.

Similar to namespaces in C#, XAML uses namespaces to define which assembly and namespace a particular class is located in. There are two ways you can do this in XAML: You can either add a namespace that directly targets a namespace from a particular assembly, or use a namespace alias in the form of a URI. The following code example shows both a namespace alias (the d namespace) and a namespace that points to a particular namespace in a particular assembly (the inf namespace).

XAML
```
xmlns:d="http://schemas.microsoft.com/expression/blend/2008"
xmlns:inf="clr-namespace:StockTraderRI.Infrastructure;assembly=
   StockTraderRI.Infrastructure"
```

The advantage of using aliases is that you can have several namespaces from several different assemblies that share the same alias. But you can also use this to have both the desktop and Silverlight assemblies and namespaces share the same XAML files.

You can add a XAML namespace in your AssemblyInfo.cs file like this:

```
C#
[assembly: XmlnsDefinition("http://URI.Goes.Here",
   "Namespace.Goes.Here")]
```

You should add the Xmlns definition attribute and Xmlns prefix attribute to both your Silverlight and desktop assemblies where they point to the correct namespace.

Once you have created a XAML configuration file, you then have to get your Silverlight application to pick up and load the configuration file. This process is often referred to as bootstrapping your application.

XAML Translation Considerations

There are a couple of things you need to keep in mind when performing the XAML Generation.

Mapping Types between .NET Framework Types and Silverlight

- When you are creating an application configuration file to be translated to XAML, it's likely that your configuration file contains .NET Framework types. For example, if you are creating an exception handler to handle an **InvalidOperationException**, you'll likely select the .NET Framework version of this exception. The XAML configuration file can only contain Silverlight types. So the .NET Framework type has to be mapped to a Silverlight type, and the mapping must include their assembly name, version, and public key token.
- For types that originate in the following assemblies, this translation happens automatically:
    - mscorlib
    - System
    - System.Core
    - System.ServiceModel
- However, if you are using types that originate from other assemblies, you will need to specify a type mapping. Only the command-line tool and the MSBuild task support this. If you can change assemblies, you can also use a XAML alias.

Silverlight 4 Types in Configuration Will Also Work in a Silverlight 5 Application

The current XAML generation tools will generate XAML configuration files that target Silverlight 4. If you are building a Silverlight 5 application, you can still use these XAML configuration files, because Silverlight will automatically use the highest version number available when instantiating a type. So if your configuration file references a Silverlight 4 file, and the client only uses a Silverlight 5 type, the type will still load normally.

Security Considerations for Using XAML Configuration Files

Do not store any sensitive information in your configuration files, such as user names or passwords. It is nearly impossible to store configuration information on the client in such a way that it cannot be accessed by the user or a system administrator.

Also, unless you transfer your configuration files over a secure connection, such as SSL, then the information in the configuration file is not protected against information disclosure. This also holds true if you embed your configuration file in the XAP, since the XAP file is essentially a zip file that will contain the configuration file as clear text.

If you like, you can implement a server-side authentication and authorization mechanism so that only authorized users are able to download your application and its configuration files. The **Configure Async** method from the **EnterpriseLibraryContainer** class contains an overload that allows you to specify credentials to use when downloading a file. However, this does not prevent authorized users from reading the configuration information.

Consuming XAML Configuration from a Silverlight Application

There are two options for consuming a XAML configuration file from your application. You can either embed it in the XAP file, or download it from a web server.

Embedding the XAML Configuration in the XAP

By default, the Enterprise Library 5.0 Silverlight Integration Pack will look in the main XAP file for a XAML configuration file called configuration.xaml. If it finds that file there embedded as a page or resource, it will attempt to load it when you first call Enterprise Library. This approach gives you almost the same experience as the .NET Framework configuration files on the desktop.

If you want to embed the configuration file in the main XAP, take the following steps:

1. Add the configuration.xaml file to your Silverlight application project.
2. Right click the configuration.xaml file and select Properties.
3. Ensure the build properties are either set to "Page" or "Resource."

Embedding the configuration file in the XAP does have a (considerable) down side. If you wish to make a change to the configuration file, you'll have to rebuild the XAP file.

Loading the XAML Configuration from the Server

It is possible to load the XAML configuration from a different location, for example from the server. This enables you to change your application's configuration without having to redeploy it.

The following code snippet shows how you can download a XAML file called ConfigurationFromServer.xaml from the server. This XAML file exists in the root of the server from which the application's XAP file was downloaded.

```
C#
EnterpriseLibraryContainer
   .EnterpriseLibraryConfigurationCompleted += (sender, e) =>
{
   if (e.ConfiguredSuccessfully)
   {
      // Continue application initialization
   }
   else
   {
      // todo: handle download or configuration error
   }
};
var configUri = new Uri("/ConfigurationFromServer.xaml",
   UriKind.Relative);
// Begin downloading the XAP file. There is also an overload
//that allows you to specify credentials to be used when
//downloading the file.
EnterpriseLibraryContainer.ConfigureAsync(configUri);
```

This sample loads a XAML file from the server, by calling **Enterprise LibraryContainer.ConfigureAsync(configUri)**. All download operations in Silverlight occur asynchronously, so this also holds true for downloading a XAML file that contains Enterprise Library's configuration. Therefore, you have to attach an event handler to the **Enterprise LibraryConfigurationCompleted** event, to continue your application initialization after Enterprise Library is loaded successfully.

If asynchronous configuration fails for some reason, for example because the connection to the server is down or the configuration file could not be found, then it's up to the developer of the application to handle the situation gracefully. Because Enterprise Library is not configured yet, the application cannot start, so you should either try to load a different XAML file or display an error message.

The following snippet demonstrates how you can configure an Enterprise Library container using a string.

```
C#
var configDictionary = (IDictionary)XamlReader.Load(
   stringWithXAMLConfiguration);
var configSource = DictionaryConfigurationSource.FromDictionary(
   configDictionary);
EnterpriseLibraryContainer.Current =
   EnterpriseLibraryContainer.CreateDefaultContainer(
      configSource);
```

Adding Enterprise Library Configuration to an Already Existing Container

It is also possible to add Enterprise Library configuration to an already existing instance of Unity. The following code sample demonstrates how you can do this:

```
C#
IUnityContainer container = new UnityContainer();
// .. (Configure the container programmatically)
// Read configuration from a string
var configDictionary = (IDictionary)XamlReader.Load(
   stringWithXAMLConfiguration);
var configSource = DictionaryConfigurationSource.FromDictionary(
   configDictionary);
// Configure Unity with the configuration
var containerConfigurator = new UnityContainerConfigurator(
   container);
EnterpriseLibraryContainer.ConfigureContainer(
   containerConfigurator,configSource);
```

If Enterprise Library could not be configured, the Exception Handling Application Block will also not work. So you cannot call the Exception Handling block or log the error using the Logging block, but you should also check in your application_UnhandledException to see if Enterprise Library was configured successfully.

Circumvent the Browser Cache When Downloading XAML Configuration Files

Internally, the **EnterpriseLibraryContainer** class uses the **WebClient** class to download XAML configuration files from the server. The **WebClient** class uses the browser cache when downloading files. While this can improve the startup performance of your application, it can also cause issues if you have modified the configuration file on the server.

If you want to make sure that you always download the latest version of the configuration file, you should employ a cache-busting technique. For example, you can append to the URL a query string parameter that changes for each download request.

In Stock Trader V2, we're appending a version parameter in the query string to each request to download a configuration file. This ensures that when the configuration and this parameter are updated, each download request gets a new URI and therefore will not use a previously cached version of the configuration file.

Configuration in the Stock Trader V2 Reference Implementation

In this section, we'll discuss how the StockTrader V2 application uses configuration files.

Requirements

These were the requirements for configuration:

- It should be possible to update Silverlight XAML configuration files without having to change the XAP files. This means that the XAML configuration files should be downloaded from the server and not embedded in the XAP.
- There are some validation rules that should be defined in configuration. These validation rules should be shared between the web services and the Silverlight application.

We really needed the ability to be able to change an application's configuration without having to recompile it. It happens quite often that we need to add or change some of the validation rules that are defined in configuration. I also sometimes need to change the logging settings to diagnose problems. It's not very practical if I need a developer to compile a new version for me every time this happens.

Configuration Architecture

Both the web services layer and the Silverlight application use Enterprise Library for configuration. The following diagram shows how Enterprise Library is configured both on the web server and in Silverlight.

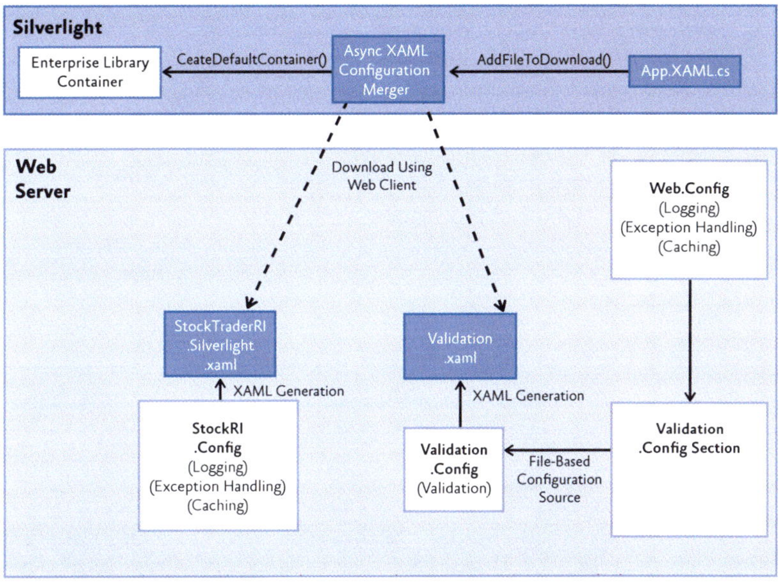

Web.config contains the configuration for the web server. For example, it contains configuration for the Exception Handling block, which is set to log all errors to the event log. The web.config file uses a file-based configuration source to reference the validation rules in a separate configuration file.

The Silverlight Stock Trader V2 application uses two separate XAML configuration files. The first file, StockTraderRI.Silverlight.xaml, contains the main configuration for the Logging, Exception Handling, and Caching blocks. The second file, validation.xaml, contains the validation rules that should be shared by both the Silverlight application and the web server.

The XAML configuration files have been created by first creating regular application configuration files (StockTraderRI.Config and validation.config) and then using the wizard to generate the XAML files.

The Enterprise Library 5.0 Silverlight Integration Pack has out-of-the box support for reading a single XAML configuration file from the XAP or downloading a single XAML configuration file from the server. For Stock Trader V2, we needed to download multiple configuration files and merge them in order to support the sharing of validation rules between the client and the server.

A common approach to merging two XAML files is to use a merge dictionary. Unfortunately, the Enterprise Library 5.0 Silverlight Integration Pack cannot use a resource dictionary or merged dictionary to store configuration information. A resource dictionary has an affinity with the UI thread, and that would cause issues when you use Enterprise Library from a background thread.

The Stock Trader V2 application contains a class called the **Async XAMLConfigurationMerger**. This class was built to download several XAML files and merge them, since this feature is not supported by the Enterprise Library 5.0 Silverlight Integration Pack out of the box. When all the asynchronous downloads have completed, it will proceed to merge them and use the merged result to configure Enterprise Library.

The Merger only merges the top level nodes in a configuration file. This means that the entire configuration for a particular block must reside in a single file. For example, you can store validation configuration XAML in one file and put the configuration for other Enterprise Library blocks in a separate file, but you cannot spread validation configuration across multiple files.

More Information

In this chapter, we discussed how you can configure the Enterprise Library 5.0 Silverlight Integration Pack. We discussed how you can use the configuration console to create application configuration files, which can then be converted into XAML configuration files. We also looked at several options for consuming these XAML configuration files in a Silverlight application.

For more information on creating application configuration files using the configuration console, please refer to the chapter **Using Configuration Tools** in the Enterprise Library documentation on MSDN®.

4 Diagnostic Logging and Tracing

The ability to analyze and diagnose an application while it is running is very important for maintainability. An application that doesn't provide good diagnostic information at run time often makes it very difficult to detect and solve problems. In an enterprise, it's not uncommon to have a huge ecosystem of hundreds of applications that should be monitored and maintained. That is why it is very important for an enterprise to have a consistent and reliable way to monitor the health of applications and to diagnose problems.

In this chapter, we'll discuss how to implement diagnostic logging and tracing in a Microsoft® Silverlight® application using the Logging Application Block from the Enterprise Library 5.0 Silverlight Integration Pack. We'll also describe the new trace listeners, which have been added to the Enterprise Library 5.0 Silverlight Integration Pack and discuss how diagnostic logging and tracing has been applied in the Stock Trader V2 Reference Implementation. At the end of this chapter, we'll show some examples.

Diagnostic Logging and Tracing with Enterprise Library

Enterprise Library provides a consistent API for writing log messages. The following example shows how you can create a log entry using one of the many overloads for the LogWriter class:

```
C#
logWriter.Write("Loading accountholder complete",
"AccountHolder", 0, 0,
    TraceEventType.Information);
```

The log entry produced by this statement contains information, such as the category or trace event type, that can be used to route the log entry to various trace listeners. Where and how you'd like to route your log entries can be defined in configuration using the configuration console. This screenshot shows an example of how log entries in the categories General, Exceptions and Validation are all written to the Isolated Storage Trace Listener, which creates a log file in isolated storage.

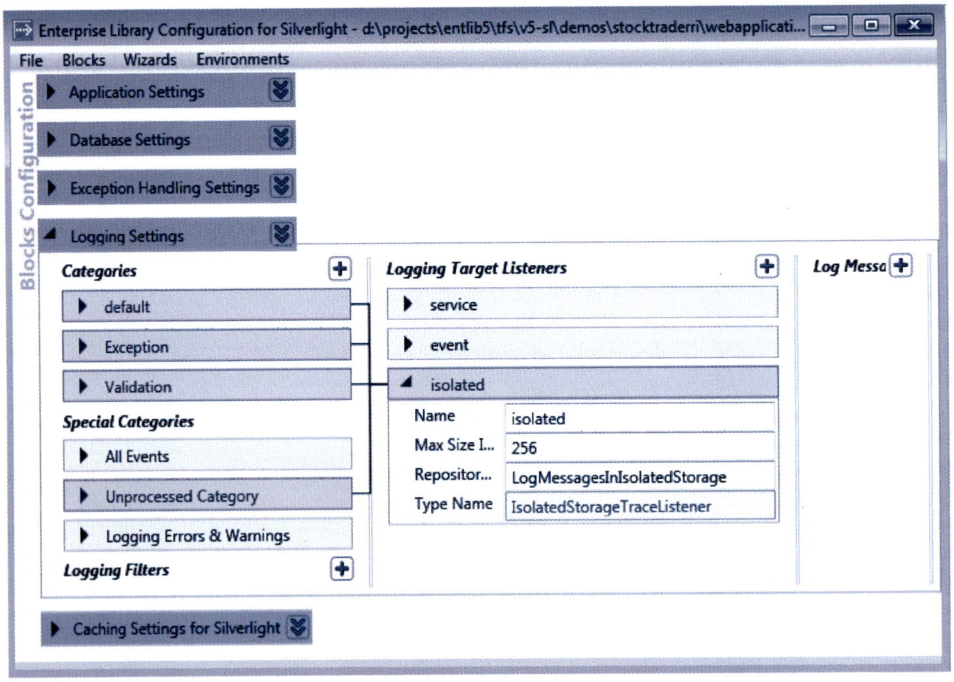

The Need for Diagnostic Logging and Tracing

While diagnostic logging and tracing are non-functional requirements for an application because the business or end-users rarely ask for this functionality, many organizations consider these requirements critical and plan them accordingly during the development cycle.

There are several reasons why you might like to implement diagnostic logging and tracing.

I like to get involved early when a new software development project starts. For me to be able to install and maintain applications, it's vital that those applications produce the right diagnostic information in a uniform way. Applications that log too little information, too much information, or log in a non-standard way make my life much more difficult. When I get involved in the development process, I can help to define stories and requirements for diagnostics. This helps to ensure that diagnostic logging and tracing is implemented as a first-class requirement, rather than an afterthought.

Detecting (Unknown) Errors

Even the most rigorous testing process cannot prevent an application from having defects when it ships. If you are lucky, your users will eventually tell you that your application has a problem. However, some users who run into problems will just abandon the application, or refuse to use it. You can use diagnostic logging and tracing to detect that your application has problems and solve them before your users start complaining.

In web applications it's very common that the web server logs any problems that happen on the web server itself. However, problems that occur on the client, for example in JavaScript, are usually ignored, resulting in a negative user experience.

While Silverlight applications usually run in the browser, they provide a much richer user experience and excellent support for handling exceptions.

Analyzing Problems

To solve a problem, you'll need to analyze and reproduce it. For straightforward bugs, this can be relatively easy. But there are also some problems that only occur in production with a particular user. When you have applied logging and tracing diligently in your application, you can get much more information about a particular problem, which will help you identify it.

Analyze Usage Patterns

Nowadays, many applications ask their users if it is ok to send anonymous information about the usage of the application.

If you log which screens are being opened, how much time is spent in a particular screen, which buttons are being pressed and which error messages are displayed, you can gain insight into how your application is being used.

This information will likely impact privacy, so you should ask the user permission first and make sure you only record anonymous information.

Diagnostic Considerations in a Silverlight Application

Diagnostic logging and tracing in a Silverlight application is somewhat different than it is, for example, for a regular Microsoft Windows® application or for a web application. The process of producing diagnostic information is largely the same, but the Silverlight security sandbox limits what additional information you can gather and where you can send the information.

I don't really miss the ability for Silverlight applications to Log to the Windows Event Log. I want to have a centralized way to monitor the health of my applications. Since Silverlight applications can be started over the Internet and can run pretty much anywhere and on many different platforms, logging to the Windows Event Log of the user is not very practical. Logging to the Windows Event Log of the web servers—now that is very useful.

No Access to Regular File System or Windows Event Log

A Silverlight application that runs in the browser doesn't have access to the regular file system. This means it's more difficult to create a log text file that can easily be found and read by your users. It is possible to log to isolated storage, though isolated storage size is limited by default to 1 MB and it is located in an obscure location. This makes it more difficult for users to discover the location and consume the file.

Another commonly used approach for regular Windows applications or web applications is to write log information to the Windows Event Log. This is not possible for Silverlight applications, because the security sandbox does not allow access to the Windows Event Log.

In the following chapters, we'll discuss how the Enterprise Library 5.0 Silverlight Integration Pack helps you to implement logging to isolated storage and automatically send the log information to a centralized location using a web service.

Not All Diagnostic Information Is Available in Silverlight

In Enterprise Library 5.0, when you create a new log message, it will also gather additional information about the host, such as the machine name, process name, process ID, and Win32® thread ID. The Silverlight security sandbox denies access to these properties, so this information cannot be gathered.

In our centralized environment for monitoring the Stock Trader application, I really wanted the ability to know which client experienced a particular problem. When a particular user experiences problems, it is very helpful to be able to find all the log messages that originated on his machine. Since Silverlight doesn't allow you to gather this information, we decided to gather it on the server. We have extended the Logging Service to also gather the IP address of the client machine.

Web Service Calls Are Asynchronous

Another difference between Silverlight and the desktop version of the .NET Framework is the fact that all Silverlight web service calls are asynchronous. While asynchronous web service calls improve the perceived performance of an application, they make it more difficult to implement certain cross cutting concerns, such as exception handling, performance measurement, and logging.

In the following sections, we'll discuss how you can apply diagnostic logging and tracing to asynchronous web services.

The Logging Application Block

The Logging Application Block will allow you to write diagnostic information. Using additional metadata, such as the severity or log categories, these log messages can be filtered and routed to various configurable trace listeners and formatters.

Most of the core for the Logging Application Block has been ported to Silverlight. However, the Trace Listeners that are provided with Enterprise Library 5.0 cannot be used in Silverlight due to the Silverlight security sandbox. For example, you cannot use the Flat File Trace Listener or the Event Log Trace Listener, because the Silverlight security sandbox denies you access to these resources. That's why the Enterprise Library 5.0 Silverlight Integration Pack ships with its own set of trace listeners, designed to be used in Silverlight.

In this chapter, we'll give you an overview of the logging application block. Then we'll discuss the Silverlight-specific scenarios that are supported by the Enterprise Library 5.0 Silverlight Integration Pack. If you would like to know more, please read the chapter on the Logging Application Block on MSDN® and in the Developer Guide.

Supported Scenarios

The Silverlight Logging Application Block supports the following scenarios:

- **Writing log messages through a consistent API.** The Logging Application Block provides a consistent API for writing log messages. That means that the same the API can be used to write diagnostic trace messages to a diagnostic trace log or important error messages to a remote web service. The **Log Writer** class has several overloads for writing log messages, with different information, such as the message, category, priority and severity.
- **Filtering and routing log messages to different trace listeners.** The logging infrastructure allows you to configure exactly which log messages should be logged and which trace listener they should be logged to.
- **Logging to a Service.** The **RemoteServiceTraceListener** provides a robust mechanism for logging certain messages to a remote server. This allows you to monitor log messages in a centralized location.

The Logging Application Block for Silverlight works great for various diagnostic logging and tracing tasks. However, you should not use it for auditing purposes. Not only is it possible for log messages to become lost if the buffer becomes full, but it is also possible to tamper with the logged messages. It is nearly impossible to implement a Silverlight auditing mechanism that cannot be circumvented. If you need auditing functionality in your application, you should implement it on the server when certain data is accessed or changed.

- **Logging to isolated storage.** The **IsolatedStorageTraceListener** allows you to write log messages in JSON format to isolated storage. This allows you to write a log file on the client computer. There is also an API for reading log messages from isolated storage, so you can display them in your application.
- **Logging to an event.** Using the Notification Trace Listener, you can receive log messages through an event. This can help you to build a developer trace window for your application.
- **Programmatic or XAML-based configuration.** The Silverlight Logging Application Block allows you to configure how you'd like to handle log messages. This configuration can be expressed in XAML files or using the programmatic API.
- **Changing log settings at run time.** When the application is running, you might want to change the log settings. For example, if a particular user is having trouble, you might want to increase the logging settings so that more information is logged. Or, if you request more space in isolated storage, you might want to increase the size that the application consumes in isolated storage.

Design of the Logging Application Block

This section gives a short introduction to the Logging Application Block Infrastructure. Most the core of the Silverlight Logging Application Block is the same as the Logging Application Block from Enterprise Library 5.0, so please refer to the Logging Application Block documentation on MSDN for more information.

The following diagram shows how the core logging infrastructure works:

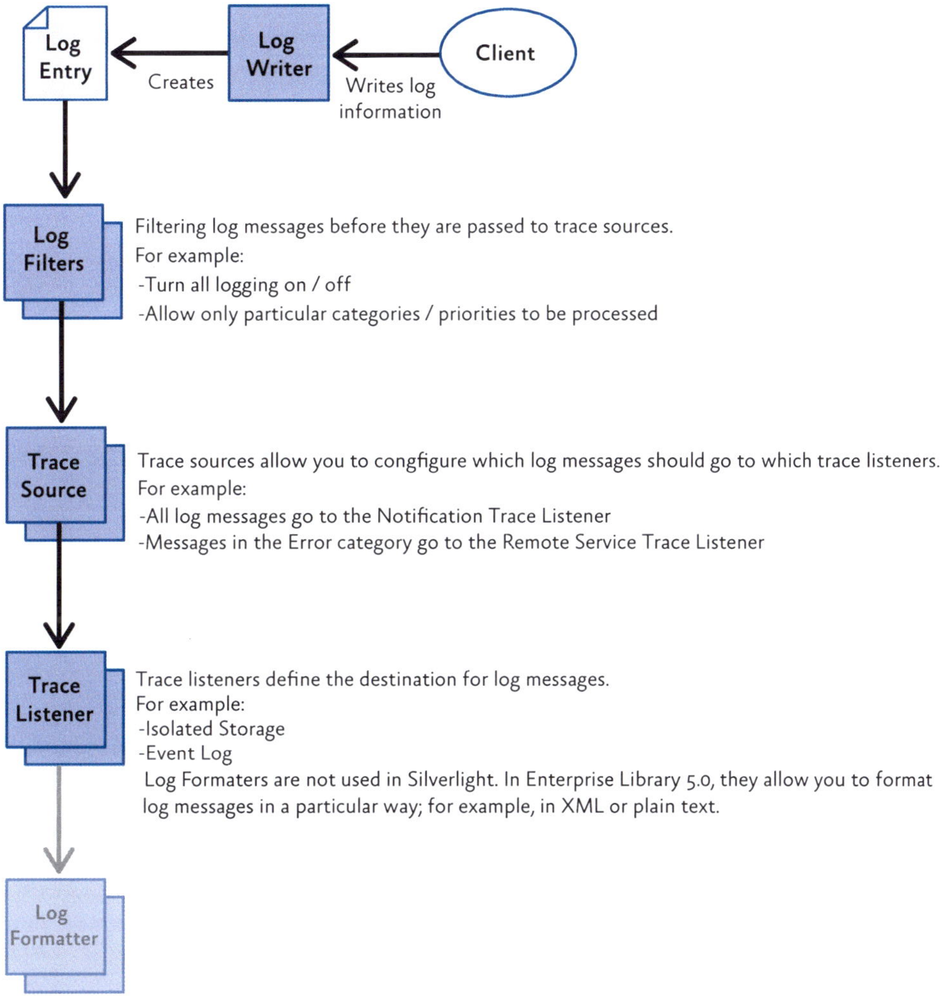

The main types of objects are:

- **Log Writer**. The log writer is the main entry point for creating log entries and writing them to your chosen logging targets. It creates an instance of a log entry containing the information to be logged, and interacts with the other objects that filter the log entry, assign it to one or more categories, format it, and dispatch it to the appropriate targets.

- **Log Entry.** Your application submits information that the Logging Application Block should log either by using the LogWriter class to automatically create a log entry or by directly creating an instance of the **LogEntry** class and populating it with the information to log.
- **Log Filters.** Log filters can block or allow a log entry based on a number of features. Each log entry is assigned to one or more categories (the default is the General category), and the category log filter can use these categories to pass or block a log entry. In addition, two special log filters can block all logging, or block log entries with a priority lower than a specified value. You define the categories, priorities, and the settings for the log filters in the configuration for the block.
- **Trace Sources.** Trace sources are effectively a set of buckets into which the block places all log entries that have not been blocked by a log filter. You use these buckets to define where log entries will be dispatched to—you can think of them as being the source of the log entries that will actually be dispatched to the target destinations. There are two basic types of trace sources:
  - There is a trace source for each category you define in the configuration of the block. These are called Category Sources.
  - There are three built-in trace sources: one that receives all log entries, one that receives log entries when an error occurs during processing or dispatching of the log entry, and one that receives all log entries that do not match any configured category. These are called Special Sources.
- **Trace Listeners.** Trace listeners represent the targets for your log entries, and you configure one for each type of target (such as a remote logging service or a log file in isolated storage) to which you want to send the log entries. Trace listeners listen for log entries arriving in the trace source buckets, format each log entry as required, and dispatch it to the target configured for that trace source. Your configuration maps each trace source (each category source you define plus the three special sources) to one or more trace listeners. The important point to note here is that this allows you to dispatch each log entry to zero, one, or more targets (such as sending it as an email message as well as writing it to the Windows Event Log). Also, multiple Trace Listeners of the same type can be used.

- **Log Formatters**. In Enterprise Library 5.0 there are several trace listeners that require log messages to be represented in text format. Enterprise Library uses Log Formatters to format the log messages into a string. For example, the log messages can be formatted as text or as XML. This feature has not been included in the Enterprise Library 5.0 Silverlight Integration Pack because its trace listeners can use the log message as is, without having it formatted. For example, the Remote Service Trace Listener will serialize and send the entire log message to the server, so it does not need to be formatted as text.

Differences between Desktop and Silverlight Versions

As mentioned before, most of the core of the Enterprise Library 5.0 Logging Application Block has been ported to Silverlight. The following features have not been ported:

- **Log Formatters.** The trace listeners in the Silverlight Logging Application Block do not use log formatters. The classes related to the log formatters have not been ported to Silverlight, to reduce the size of the Silverlight Logging Application Block assembly.
- **Instrumentation.** Enterprise Library 5.0 implements instrumentation in the logging block, for example, to provide performance counters. This feature has not been ported, since performance counters are not available in Silverlight.
- **Revert impersonation property.** Enterprise Library 5.0 uses impersonation when logging to improve logging performance. Impersonation itself is not supported in Silverlight, so the **Revert Impersonation** property does nothing.
- **Limited supported types for extended properties** . The LogEntry class allows you to add objects as extended properties, but the trace listeners from the Enterprise Library 5.0 Silverlight Integration Pack do not support using arbitrary objects. Since the isolated storage trace listener uses the DataContractJsonSerializer, it only supports primitive types. The RemoteServiceTraceListener is using a data contract (LogEntryMessage), which only supports strings as extended properties.
- **Stack Trace**. In Enterprise Library 5.0, it's really easy to write the stack trace when a LogEntry is being written. The stack trace is not part of the **LogEntry,** but at the time a log message is being formatted, the log formatters can simply grab the stack trace

If you need a stack trace for each log entry, then you could create a derived version of the trace listener you'd like to use and add the stack trace to the log message as an extended property.

from the context and add that to the formatted log entry. Since the Enterprise Library 5.0 Silverlight Integration Pack doesn't support log formatters, you will also not automatically get stack trace information for each log entry.

Remote Service Trace Listener

With the Remote Service Trace Listener, it is possible to log events to a Windows Communication Foundation (WCF) service. The Enterprise Library 5.0 Silverlight Integration Pack provides both the trace listener to be used on the Silverlight client and a configurable logging service implementation to be used on the server. The Remote Service trace listener can send messages immediately or in a batch. It is also robust enough to handle scenarios in which logging to the service fails. In those cases, the log entries can be buffered to memory and isolated storage. This means that log entries that could not be sent can be sent later when the application restarts.

The following diagram shows how the Remote Service trace listener works:

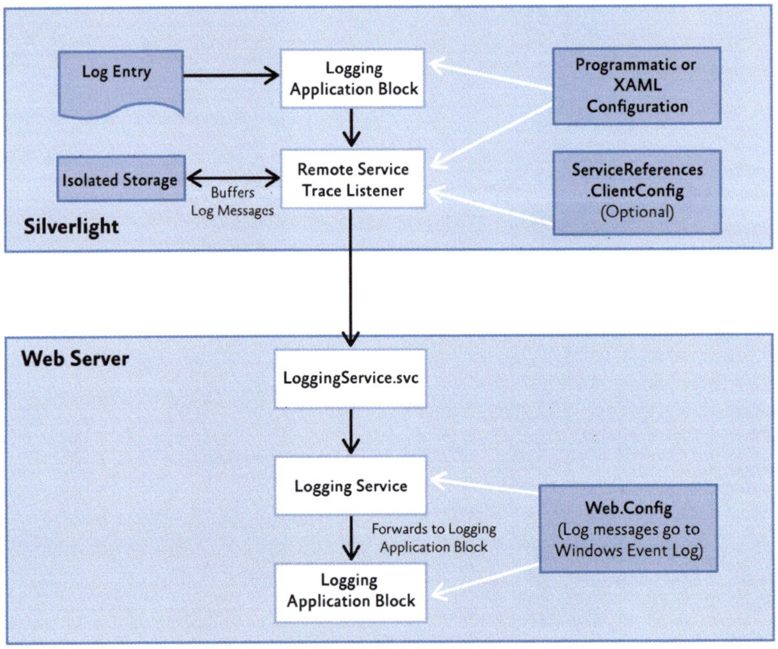

You can configure the Logging Application Block to send certain log entries to the Remote Service trace listener. The Remote Service trace listener can either send the log entries immediately, or as a batch. If the log entries cannot be sent, then they will be buffered in memory and optionally to isolated storage. You will need to configure how the Remote Service trace listener should contact the server, by specifying the address and binding to use. This can be done either programmatically or using a ServiceReferences.ClientConfig file.

On the server, the log messages are received by the log service. This log service needs a service (.svc) file to define the location of the endpoint. The binding for the service also needs to be configured in the web.config file. The log service forwards the log entries to the Enterprise Library 5.0 Logging Application Block.

Parameters

The Remote Service trace listener can be configured with the following parameters:

- **Logging Service Factory**
  The Remote Service trace listener needs to know how and where to send the log messages. Typically, you'll set this value to the name of the endpoint, which will create and configure a Logging Service Factory for you, but you can also set it to a custom Logging Service Factory.
- **Send Immediately flag, and Submit Interval**
  You can specify whether the service should send log messages immediately, or if it should send items in a batch. The submit interval determines how often batched messages should be sent, if they are not sent immediately.
- **Buffer size**
  Log messages can be buffered in memory by specifying a value for **MaxElementsInBuffer**, and to isolated storage by specifying a value for **IsolatedStorageBufferMaxSizeinKilobytes**. This buffering is used when messages should be sent in a batch, or when log messages cannot be sent for some reason (for example when the Internet connection is lost). You can specify the maximum number of items to buffer. When you also specify a maximum size for the isolated storage buffer, then this trace listener will also attempt to store the maximum number of items in isolated storage. If the isolated storage buffer becomes full, older messages will be overwritten.

The minimum size for **IsolatedStorage BufferMaxSizein Kilobytes** is 5 Kb. Setting it to a value that is lower than this would not make sense, because the buffer would be too small to effectively store items in it, but you would still incur a significant performance hit for serializing.

Logging Service Factory

- The Remote Service trace listener uses WCF to send log messages to the server. Like any WCF service, you will need to configure the address and binding of the service. There are two ways in which you can configure these:
- **Configure the binding in a ServiceReference.ClientConfig file.** The most common way to configure a WCF service is to add a client config file. This file describes the name of the endpoint and its address and binding. You can configure the Remote Service trace listener by setting the value of the Logging Service Factory property to the name of the binding.
- **Supplying a custom Logging Service Factory.** If you wish to have more control over the way the WCF service is created, you can also supply an instance of a Logging Service Factory to the Logging Service Factory property. For example, if you wish to secure your logging service, then you can use a custom Logging Service Factory to supply credentials to the Logging Service. You can only supply a custom Logging Service Factory directly in a XAML configuration file or using the fluent configuration API. It cannot be set using the configuration console or added to an application configuration file.

- The section entitled **Using the Remote Trace Listener** describes how you can use these settings in more detail.

Buffering Messages in Memory

If you specify that messages should be buffered, they will always be buffered in memory. You can also specify that the log messages should be buffered to isolated storage, but the messages will still be buffered in memory.

Buffering to Isolated Storage

- When you use the option to buffer log messages in isolated storage, the Remote Service trace listener will immediately try to claim the size of the buffer when it starts. If there is not enough space available, then it will claim the amount of space that is actually available. Once claimed, the buffer size will not be automatically changed, even if you change the size in the XAML configuration. However, using the Runtime Changes API, which is described later in this chapter, you can manually change the size of the buffer.
- The isolated storage buffer will not exceed the maximum buffer size. When the buffer becomes full, it will roll over and start to overwrite the oldest log messages. If you attempt to add an item that by itself exceeds the size of the buffer, then it will not be added to the isolated storage buffer. No logging error will be generated though, because it will still be added to the in-memory buffer.
- When you start multiple instances of your application, only the first instance will be able to use the isolated storage buffer. The other instances will only be able to buffer in memory.

Be careful with specifying a huge buffer. If the message that is sent to the server contains too many log messages and exceeds the maximum message size specified by WCF, it will not be able to send log messages again and will retry indefinitely. Of course, you can solve this by increasing the maximum message size or decreasing the buffer size. But you can also create a class that derives from the Remote-ServiceTraceListener and send the log messages in smaller chunks. In that case, don't forget to create a custom RemoteService TraceListenerData class as well.

Configuring the Logging Endpoint on the Server

On the server side, the logging service can be used as is, customized to fit a particular scenario, or replaced entirely by a custom implementation. The LogService.svc file defines where the endpoint for your logging service resides. In this file, you can specify which implementation you'd like to use. The default logging implementation is the **Logging Service** in the **Microsoft.Practices.EnterpriseLibrary.Logging.Service** namespace. You can also use this class as a base class. There are two virtual methods you can use to customize the logging process: **Translate**, which creates a log entry from the incoming **LogEntryMessage**, and **CollectInformation**, which allows you to add additional information to the log message.

The Logging Service uses the Logging Application Block from Enterprise Library 5.0 to handle log messages on the server. This means you can use all the filter options, trace listeners, and log formatters that are available. For example, you can log all warnings and errors to the Windows Event Log using the Event Log trace listener, and log all informational log messages to a rolling flat file using the Flat File Trace Listener.

Since the Logging Service is implemented as a WCF service, you'll also need to configure the endpoint in the web.config file.

I wanted a way to correlate log entries with the user's machine. If a particular user has problems, we can increase his logging levels, but I would like a way to filter log messages for just this user. We implemented this in Stock Trader V2 by querying the **HttpContext.Current.Request.UserHostName** property. This works fine for the purpose of diagnosing problems. However, since this value can be spoofed by a malicious client, you should never use something like this for security-critical actions such as auditing.

Isolated Storage Trace Listener

The **Isolated Storage Trace Listener** allows you to write log messages in JavaScript object notation (JSON) format to isolated storage. This allows you to create a log file on the client computer. The following diagram shows how the Isolated Storage trace listener works:

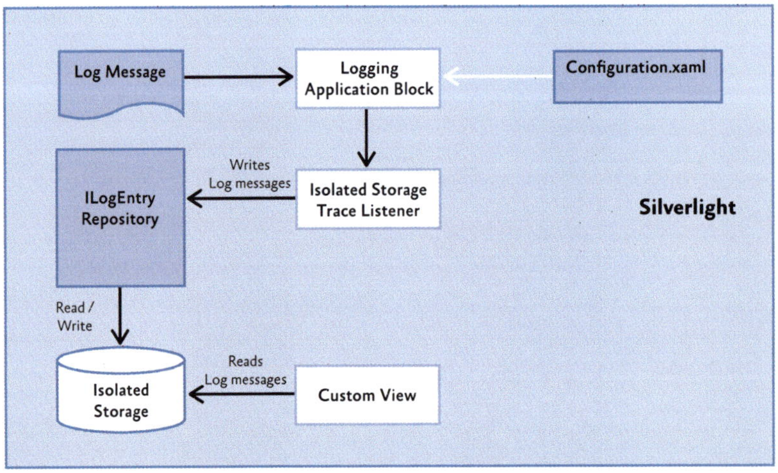

The Isolated Log trace listener has the following properties you can set, either programmatically or through configuration:

- **Name**
  This is the name that will be used to create the log file in isolated storage. Only a single Isolated Storage trace listener instance can write to a file with a particular name at the same time, even across application instances.
- **Maximum Size in Kilobytes**
  You can specify the maximum size the log will use. If you specify a size that is larger than the actual available size, it will only use the actual available size.

Maximum Size

- When the Isolated Storage trace listener starts, it will immediately try to claim the specified size. If there is not enough space available, it will claim the amount of space that is actually available. Once claimed, the isolated storage log size will not be automatically changed, even if you change the size in the XAML configuration. However, using the runtime changes API, which is described later in this chapter, you can manually change the size of the buffer.
- When the log becomes full, it will roll over and start to overwrite the oldest log messages. If you attempt to add an item that by itself exceeds the size of the log, then it will not be added to the log file at all and a logging error will be generated.

Reading Log Messages from Isolated Storage

If you want your users to be able to view the log file, you should create a custom view for this purpose. You can use the **ILogEntry Repository** interface to retrieve log messages from the isolated storage log.

Notification Trace Listener

The Notification Trace Listener raises an event whenever a new log entry is received. On this event, you can add custom logic. One possible usage would be to store all log messages in memory, and display them in a custom view.

Runtime Changes API

The Silverlight Logging Application Block provides an API for making changes to the logging settings at run time. Note that changes made to the logging settings through this API are not persisted. When the application restarts, it will revert back to the settings defined in configuration. This API allows you to do the following:

- **Read current trace listeners and trace sources.** You can use the API to find the currently configured trace listeners and trace sources so that you can retrieve their settings.
- **Enable or disable logging altogether.** The API allows you to turn logging on or off altogether.
- **Changing existing trace listeners.** You can change the settings for an existing trace listener. For example, for the Isolated Storage trace listener you can change the size of the isolated storage cache, or for the Remote Service trace listener, you can change the interval at which log messages are being sent to the server and the buffering settings.
- **Changing existing trace sources.** You can also change existing trace sources. For example, you can change the source level of a trace source. You can also add or remove trace listeners that have already been configured.

It is not possible to add new trace sources or trace listeners. Any trace listener you'd like to add to a trace source using this API should already be used by another trace source.

You can't define a trace listener in configuration that is not being used, because it will be ignored when the Logging Application Block is configured. If you have a trace listener that's normally not used, but you'd like to be able to add it using the change API, then you can do the following: define an unused trace source in configuration and add the trace listener to it. That way, it's available when the Logging Application Block is configured and you can then use the change API to add it to a different trace source.

Logging and Tracing in the Stock Trader V2

It is time to discuss how diagnostic logging and tracing were implemented in the StockTrader V2 reference implementation. The first version of the Stock Trader didn't have very good diagnostic logging and tracing functionality. This caused a number of unfortunate side effects:

- The IT organization usually didn't know about problems with the old Stock Trader until unhappy users reported them.
- Problems reported by users were often difficult to reproduce. Only a very limited amount of diagnostic data was generated by the application.
- When a power user ran into a problem with the old Stock Trader, he couldn't get access to any diagnostic information either. Given more information, these users are sometimes able to solve issues by themselves (if the problem was on their own computer) or at least be able to help the IT organization analyze the problem.

Stories for Logging and Tracing

In this section, we'll discuss the most important stories for logging and tracing, and how they were implemented in Stock Trader V2.

When we began designing Stock Trader V2, we didn't want logging and tracing to become an afterthought. Rather we wanted to make it a first-class requirement in our planning. To do this, we defined several stories around diagnostic logging and tracing, with different actors, such as a user, a developer, or somebody from IT operations. These stories were planned, developed and (acceptance) tested along with the other functional stories for Stock Trader V2.

There are many ways in which you can realize a centralized method for recording and viewing problems that occur within an application. For example, you can log to a central database and build an application that monitors this database.

In our organization, we already use System Center Operations Manager to monitor the Event Logs of our servers. When a problem occurs, such as an unhandled exception or warning, our IT staff is notified of this problem and we can start to investigate it. We also wanted this capability for the Silverlight front end of Stock Trader V2.

IT Operations Should Be Able to View Errors in a Centralized Way

IT operations needs a centralized way to record, detect, and view problems that occur with the entire Stock Trader V2 application. This means that problems that occur within the Silverlight client and problems that occur only within the server should be recorded in a central location.

Errors and warnings that occur in the Silverlight client need to be sent back to the web servers, where they should be added to the Windows Event Log. From there, they can be picked up by System Center Operations Manager.

The following diagram illustrates how warnings, errors, and critical messages are recorded to the Windows Event Log of the server:

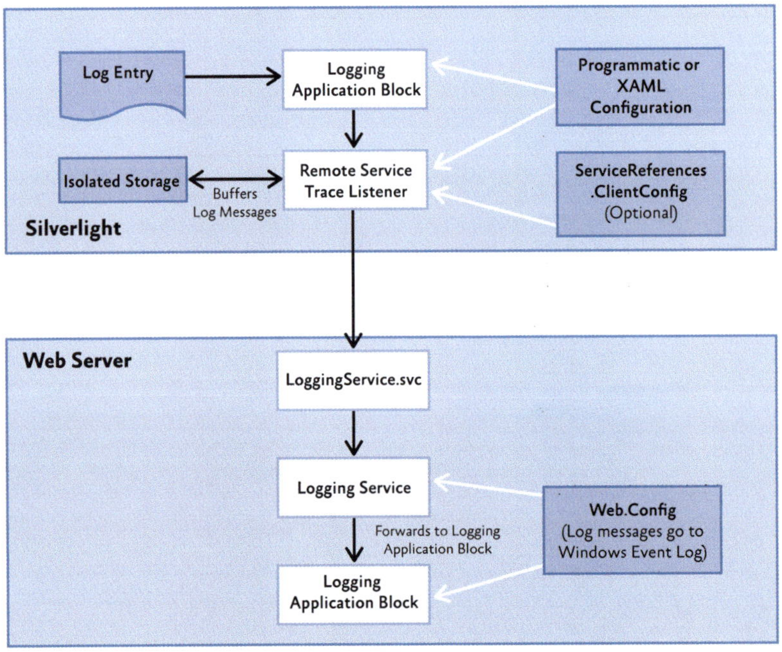

In Stock Trader V2, exception handling is performed by the Exception Handling Application Block. This block is configured so that certain types of exceptions are logged using the Logging Application Block, with an appropriate log severity. There are also many explicit calls to the Log Writer in the Stock Trader. For example, log messages are written when the user clicks a button, when a validation error occurs, and when a web service successfully completes.

In Stock Trader V2's configuration file (StockTraderRI.Configuration.xaml), the logging application block is configured so that log messages with a severity of Warning, Error, or Critical are sent to the

Remote Service trace listener. This listener will send log messages to the log service on the web server.

To prevent log messages from getting lost if there are connection issues, log messages will also be buffered in isolated storage. This way the log messages will even be sent if the application has crashed completely and is restarted.

On the web server, any log messages that are received by the log service will be sent to the Logging Application Block, which is configured to send all log messages with a severity of Warning, Error, or Critical to the event log.

Power Users Should Be Able to See Critical Logging Information

Some of the users of Stock Trader V2 are advanced computer users. They would like to be able to see the information that's logged by Stock Trader V2. When a problem occurs in the application, they want to see what exactly is going wrong. Sometimes, they are able to solve the issue. If not, they are able to provide the IT organization with additional information.

The following diagram shows how this is solved in Stock Trader V2:

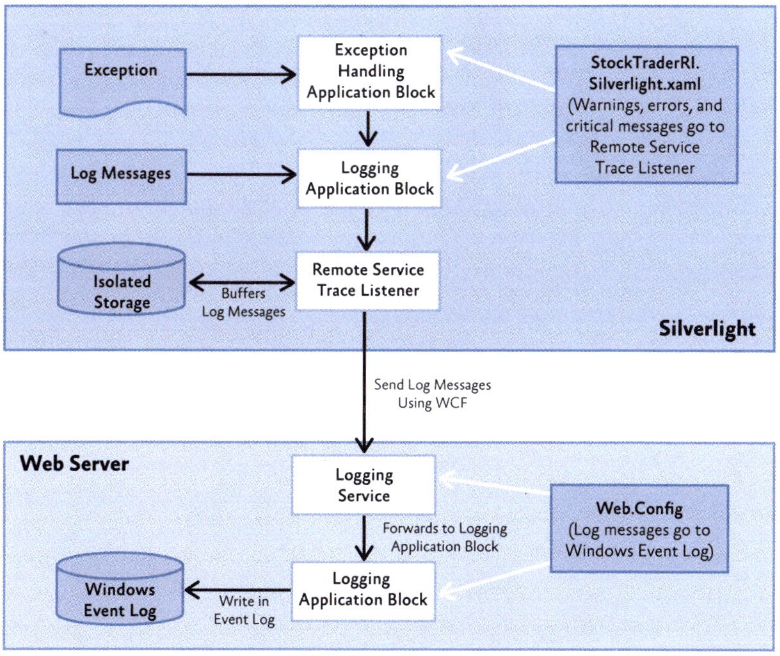

In the configuration file (StockTraderRI.Silverlight.xaml) it's defined that log messages with a severity of Warning, Error, or Critical will be stored in isolated storage, by using the Isolated Storage Trace Listener. A custom view, called the **Isolated Storage Log View**, was built to read and display these log messages from isolated storage. It can read log messages from the Isolated Storage Log Entry Repository. Since this feature is only available for power users, we decided that it doesn't need a menu item, but can be accessed when you add **#Isolated StorageLog** to the URL in the browser. In this view, the users should be able to copy the contents of a log entry to the clipboard so that it can easily be sent to IT operations.

The isolated storage is by default only 1 Mb in size. Therefore Stock Trader V2 at startup asks for more isolated storage space. If the user approves this request, then the size of the log file is increased.

Users Should Be Able to Increase Their Logging Levels at Run Time

When a user has a particular problem in production that is difficult to reproduce, it should be possible for that user to increase the logging levels of the application at run time. This means that more log diagnostic information will be sent back to the server. This can help to diagnose a problem with a particular user.

Developers Should Be Able to See Detailed Tracing Information

While developing or maintaining Stock Trader V2, developers need the ability to see more detailed tracing information. This information should be visualized as the application is running. If a problem occurs, this information can help the developer to more quickly understand a particular problem.

The following diagram shows how this was implemented in Stock Trader V2:

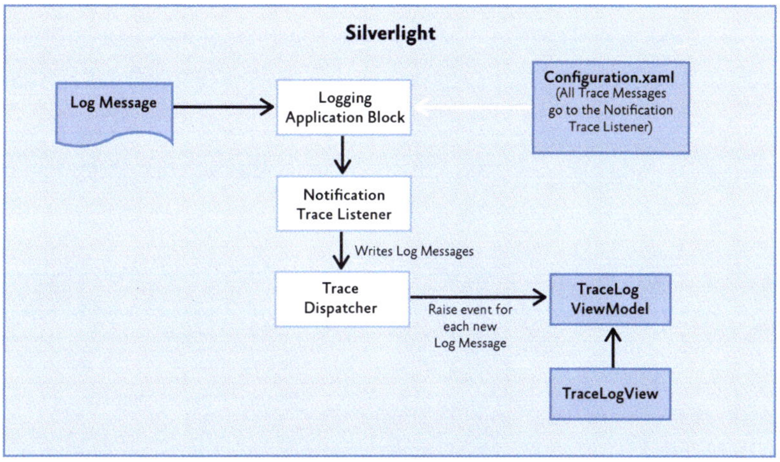

In the Stock Trader V2's configuration file, the All Events Trace Source has been set to use the Notification trace listener. This means that all log messages are sent to this trace listener. When a message is logged, the trace dispatcher will raise an event. The Trace Log View Model listens for these events and will display all received messages.

Being Able to Correlate Log Messages with Activities

Developers and IT professionals would like the ability to group log messages by activity or category. When a problem occurs while executing a particular activity, they can easily see which other log messages were also part of that same logical activity. For example:

- Which log entries were the direct result of a particular button click?
- Which log entries were generated on a particular screen?

When you create a tracer, any log message that is written until that tracer is disposed will be associated with that tracer. A tracer has both an activity ID and an associated category name, so all log messages that will be written during the lifetime of that tracer will also get that activity ID and category name associated. For more information about creating tracers, please refer to the chapter on Creating Tracers.

Creating a tracer for synchronous operations is fairly straightforward. However, for asynchronous actions, you'll have to take additional steps to have the tracing context transferred to the new thread. Just before you start a new asynchronous action, you'll need to cap-

ture the tracing context in a variable. Then, on the new thread, you can restore the tracing context. Since all web service actions are asynchronous in Silverlight, there are many circumstances in which you might run into this situation.

Consider the following example. You want to load a particular account holder (let's say, account holder 621). Before you call the load account holder web service, a log message is written that you are about to load account holder 621. When the web service call returns, it throws an exception; for example: "Account holder could not be found." This exception is also logged, but since it does not mention which account holder could not be found, it can be difficult to debug. With a synchronous action, you can easily use a tracer to correlate these two log messages. With asynchronous actions, you can also use tracers, though you will need to take additional steps.

The Silverlight Logging Application Block allows you to capture the tracing context into an object, just before you start an asynchronous action. Most asynchronous actions allow you to pass an object as user state. This object is then returned when the asynchronous action completes. You can then pass the captured context to the asynchronous action as the user state and use it to restore the captured tracing context when the request completes.

The following code block shows a slightly simplified example of how this is implemented in Stock Trader V2.

C#

```
public void LoadAccountHolder()
{
   // Create a tracer that groups all log messages together
   using (logWriter.StartTrace("LoadAccountHolder",
      Guid.NewGuid()))
   {
      // Capture the context just before
      //the async action starts
      var capturedContext = Tracer.CaptureTracingContext();
      // Pass the captured tracing context as
      //user state to the async action
      accountHolderServiceAgent.GetAccountHolder(
         LoadAccountHolderComplete, capturedContext);
   }
}

private void LoadAccountHolderComplete(
   ServiceResult<AccountHolder> serviceResult)
{
   // On the callback, retrieve the captured
```

```
    //tracing context from the user state
    var capturedContext = serviceResult.UserState as
        ICapturedTracingContext;
    // Restore the tracing context
    using (capturedContext.Restore())
    {
        // Any log message that is written here is
        //associated with the same tracing context as
        //in the LoadAccountHolder method
    }
}
```

Implementing Diagnostic Logging and Tracing in Your Application

The basics for implementing logging and tracing in your application are already described in detail in the **Enterprise Library documentation** on MSDN and in the **Developer's Guide**.

In this chapter, we'll discuss how you can implement the Silverlight-specific scenarios, such as using the Remote Service trace listener in your application.

Using the Remote Service Trace Listener

In this section, we'll describe how you can log messages to a remote service. To complete these steps, you will need a web application project that will host the WCF service. This can be the same web application that hosts the Silverlight application. You will also need a Silverlight application. These projects do not need to be together in the same solution.

When you want to use the Remote Service trace listener, you should perform the following steps:

1. Add the log service to your web server
2. Configure the Logging Application Block on the web server
3. Configure the WCF binding in the Silverlight application
4. Configure the Logging Application Block to use the Remote Service trace listener

These steps are described in more detail in the following paragraphs.

Adding the Log Service to Your Web Server

First, you will need to add the required references to your web application:

- Microsoft.Practices.EnterpriseLibrary.Common.dll
- Microsoft.Practices.EnterpriseLibrary.Logging.dll
- Microsoft.Practices.EnterpriseLibrary.Logging.Service.dll

Then you will need to create the endpoint by adding the Log Service.svc file to your solution. You can also use a different name for this file, but this name defines the URL used to access the service. This should be the contents of the LogService.SVC file:

```
SVC
<%@ ServiceHost Language="C#" Debug="true"
Service="Microsoft.Practices.EnterpriseLibrary.Logging.Service.
LoggingService" %>
```

This creates an endpoint that uses the **LoggingService** class that is located in the Microsoft.Practices.EnterpriseLibrary.Logging. Service assembly. If you wish to use a different implementation for the logging service, then you should change the Service property accordingly.

Then you will need to configure the binding to use for this service in the web.config file. The following example shows how you can set up a binding for this service that uses binary message encoding and HTTP for transport.

```
Web.Config
<configuration>
  <system.serviceModel>
    <bindings>
      <binding name=
        "Microsoft.Practices.EnterpriseLibrary.
        Logging.Service.customBinding0">
        <binaryMessageEncoding />
        <httpTransport />
      </binding>
    </bindings>
    <services>
      <service name="Microsoft.Practices.EnterpriseLibrary.
        Logging.Service.LoggingService">
        <endpoint address="" binding="customBinding"
          bindingConfiguration="Microsoft.Practices.
```

```
                EnterpriseLibrary.Logging.Service.customBinding0"
                contract="Microsoft.Practices.EnterpriseLibrary.
                Logging.Service.ILoggingService" />
         </service>
      </services>
   </system.serviceModel >
</configuration>
```

You can use any binding that is supported by Silverlight, but keep in mind that both the server and the client need to be configured the same way. This page on MSDN describes all the available bindings you can use in Silverlight.

> **Note:** *For brevity, this sample doesn't use any form of message or transport security, so it is possible to intercept or tamper with the log messages. If you are planning to log sensitive information, consider using message or transport security. Also consider storing the log messages in a secure location.*

Configuring the Logging Application Block on the Server

The Logging Service will use the Logging Application Block on the web server to process the log messages from the Silverlight client. You will need to configure the Logging Application Block to handle the log messages appropriately.

The following example shows how you can configure the Logging Application Block in the web.config file to send all the log messages to the Windows Event Log:

```
Web.config
<configuration>
   <configSections>
      <section name="loggingConfiguration" type="Microsoft.
         Practices.EnterpriseLibrary.Logging.Configuration.
         LoggingSettings,Microsoft.Practices.EnterpriseLibrary.
         Logging, Version=5.0.414.0, Culture=neutral,
         PublicKeyToken=31bf3856ad364e35"
         requirePermission="true" />
   </configSections>
   <loggingConfiguration name="" tracingEnabled="true"
      defaultCategory="General">
      <listeners>
         <add name="Event Log Listener" type="Microsoft.
            Practices.EnterpriseLibrary.Logging.TraceListeners.
            FormattedEventLogTraceListener,
```

```
            Microsoft.Practices.EnterpriseLibrary.Logging"
            listenerDataType="Microsoft.Practices.
            EnterpriseLibrary.Logging.Configuration.
            FormattedEventLogTraceListenerData,
            Microsoft.Practices.EnterpriseLibrary.Logging"
            source="Enterprise Library Logging"
            formatter="TextFormatter"
            log="" machineName="." traceOutputOptions="None" />
      </listeners>
      <formatters>
         <add type="Microsoft.Practices.EnterpriseLibrary.
            Logging.Formatters.TextFormatter,
            Microsoft.Practices.EnterpriseLibrary.Logging"
            template="Timestamp: {timestamp}{newline}&#xA;
            Message:{message}{newline}&#xA;Category: {category}
            {newline}&#xA;Priority: {priority}{newline}&#xA;
            EventId: {eventid}{newline}&#xA;Severity:
            {severity}{newline}&#xA;Title:{title}
            {newline}&#xA;Machine: {localMachine}{newline}&#xA;
            App Domain:{localAppDomain}{newline}&#xA;ProcessId:
            {localProcessId}{newline}&#xA;Process Name:
            {localProcessName}{newline}&#xA;Thread Name:
            {threadName}{newline}&#xA;Win32
            ThreadId:{win32ThreadId}{newline}&#xA;Extended
            Properties:{dictionary({key} - {value}{newline})}"
            name="Text Formatter" />
      </formatters>
      <categorySources>
         <add switchValue="All" name="General" />
      </categorySources>
      <specialSources>
         <allEvents switchValue="All" name="All Events">
            <listeners>
               <add name="Event Log Listener" />
            </listeners>
         </allEvents>
      </specialSources>
   </loggingConfiguration>
</configuration>
```

For more information on how to configure the Logging Application Block, please refer to the chapter on **Configuring the Logging Application Block** on MSDN or the **Developer's Guide**.

Configuring the WCF Binding in the Silverlight Application

Now that the server is configured, we will also need to configure the WCF binding on the client. To do this, you will need to add an XML file called ServiceReferences.ClientConfig to your web application. Set the build action for this file to "Content," so that it ends up in the XAP file for your application.

In the ServiceReferences.ClientConfig file, you will configure what binding the Logging Service should use. The following is an example of a ServiceReferences.ClientConfig file:

ServiceReferences.ClientConfig

```
<configuration>
  <system.serviceModel>
    <bindings>
      <customBinding>
        <binding name="CustomBinding_ILoggingService">
          <binaryMessageEncoding />
          <httpTransport maxReceivedMessageSize="2147483647"
            maxBufferSize="2147483647" />
        </binding>
      </customBinding>
    </bindings>
    <client>
      <endpoint address="/LoggingService.svc"
        binding="customBinding" bindingConfiguration=
        "CustomBinding_ILoggingService"
        contract="Microsoft.Practices.EnterpriseLibrary.
        Logging.Service.ILoggingService"
        name="CustomBinding_ILoggingService" />
    </client>
  </system.serviceModel>
</configuration>
```

The binding configuration in this file should match the binding configuration for the server. In this case, we're using binaryMessage Encoding over HTTP.

A couple of things are very important in this file:

- The **address** property on the **endpoint** element refers to the location where your service endpoint is located. In this case, it assumes that the service endpoint is located on the same server where the XAP file was deployed. If the logging service is located in a different location, of if you have given your SVC file a different name, you should change this URL.

- The **name** property on the **endpoint** element will be used in the next step, where we're configuring the Logging Application Block to use the Remote Service trace listener.

For more information on creating ServiceReference.ClientConfig files, please refer to the following page on MSDN.

Configure the Logging Application Block to Use the Remote Service Trace Listener

Now we'll create a XAML configuration file that will send all events to the Remote Service trace listener.

1. Open the Enterprise Library configuration console.
2. Select **Blocks -> Add Logging Settings for Silverlight**

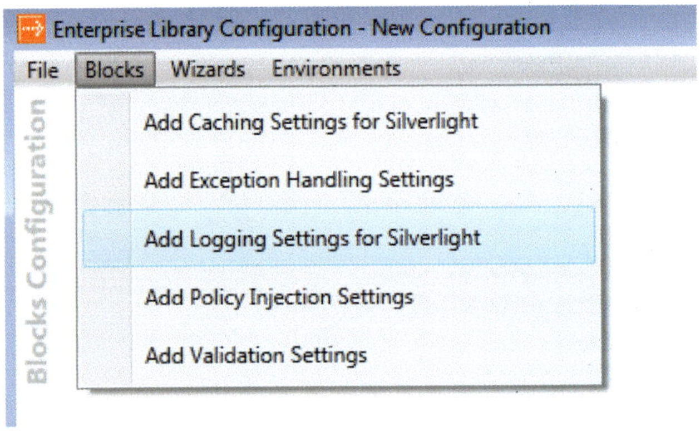

This automatically adds a Remote Service trace listener to the configuration.

3. Open the **All Events** special category and click the plus sign (+) to add the Remote Service trace listener.

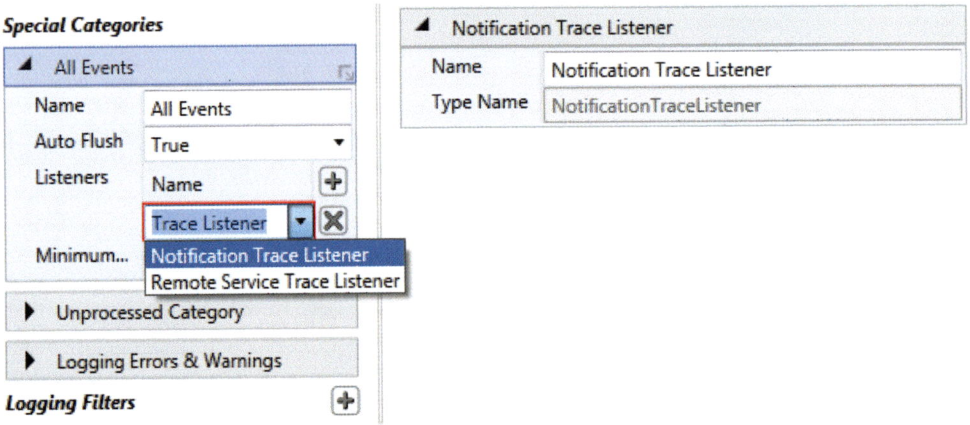

4. Open the Remote Service trace listener and set the value of the **Logging Service Factory** property to the name of the binding that was configured in the previous step. In this example, this is **CustomBinding_IloggingService**.
5. Select **Wizards -> Export to XAML**. Browse to the root of your Silverlight project and save the file there under the name Configuration.xaml.

6. Add the Configuration.xaml file to your Silverlight application project and set the build action to **Resource** or **Page** in the file properties.

Your Silverlight application is now configured to send all log messages to the Remote Service trace listener. On the server, all the log messages from the Silverlight client will be written to the Windows Event Log.

Using a LogServiceFactory to Configure the WCF Binding Programmatically

Instead of using a ServiceReferences.ClientConfig file and specifying the name of the binding in the configuration.xaml file, you can also configure the binding for the Logging Service programmatically. You will do this by specifying a custom Logging Service Factory for the Remote Service trace listener.

> **Note:** *You can only configure a custom Logging Service Factory in code or a XAML configuration file. You cannot select a LogServiceFactory using the configuration console, nor can you set it in an application configuration file and export it to a XAML configuration file.*

You will first need to create a custom LoggingServiceFactory that creates the channel programmatically. The following example shows how to create a custom Logging Service Factory that sets up the binding to use Binary Message Encoding and Transport over HTTP.

Customizing the Log Service by Creating a Derived Class

If you wish to customize the Logging Service, for example to add additional information to the log messages, then you can create a class that derives from the Logging Service. By overriding the **Collect Information** method, you can add additional information to the log message. For example, you can add the IP address of the caller to the log message.

The following example shows an example of an extended logging service.

```
C#
public class ExtendedLoggingService : LoggingService
{
   protected override void CollectInformation(Microsoft.
   Practices.EnterpriseLibrary.Logging.LogEntry entry)
   {
      // Get the user IP address and add it as an
      // extended property. NOTE. This information can easily
      // be spoofed by the caller. Use this information only
      // for diagnostic purposes, not for authentication or
      // auditing purposes.
      var messageProperties = OperationContext.Current.
         IncomingMessageProperties;
      var endpoint = messageProperties[
```

```
            RemoteEndpointMessageProperty.Name] as
            RemoteEndpointMessageProperty;
        string clientAddress = endpoint != null ?
            endpoint.Address : string.Empty;
        entry.ExtendedProperties.Add("ClientIPAddress",
            clientAddress);
        // Also add the client host address as a category. This
        // allows you to turn on server side logging for a
        // single client
        if (!string.IsNullOrEmpty(clientAddress))
        {
            entry.Categories.Add(clientAddress);
        }
        base.CollectInformation(entry);
    }
}
```

To use this, you will need to update the LoggingService.svc file to point to the new extended service:

```
LoggingService.svc
<%@ ServiceHost Language="C#" Debug="true"
Service="StockTraderRI.Logging.ExtendedLoggingService" %>
```

And you will need to update the web.config to also use the Extended Logging Service:

```
Web.config
<service name="StockTraderRI.Logging.ExtendedLoggingService">
    <endpoint address="" binding="customBinding"
        bindingConfiguration="Microsoft.Practices.
        EnterpriseLibrary.Logging.Service.customBinding0"
        contract="Microsoft.Practices.EnterpriseLibrary.
        Logging.Service.ILoggingService" />
    <endpoint address="mex" binding="mexHttpBinding"
        contract="IMetadataExchange" />
</service>
```

Using the Isolated Storage Trace Listener

In this section, we'll describe the steps you should take to log all messages to the Isolated Storage trace listener. We'll also describe how you can read the written log messages.

Configuring the Isolated Storage Trace Listener

1. Open the Enterprise Library configuration console.
2. Select **Blocks -> Add Logging Settings for Silverlight**

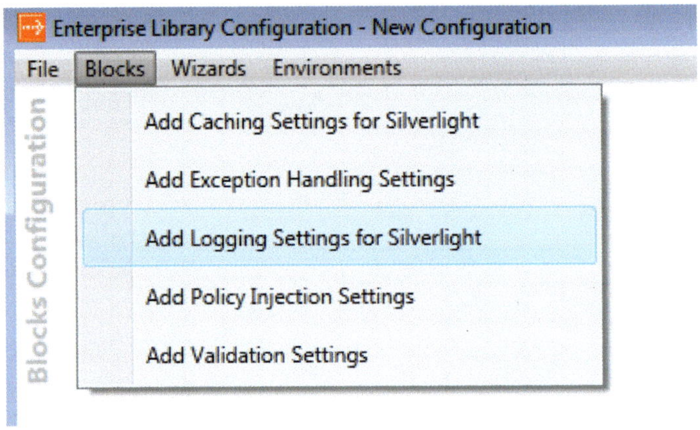

3. Add the Isolated Storage trace listener.

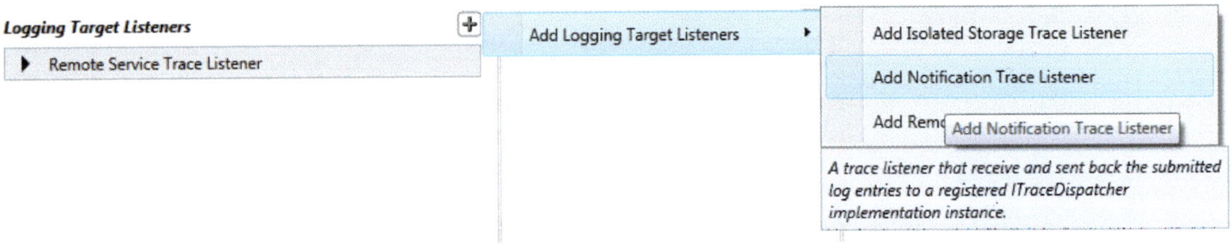

4. Specify the name of the repository. This name is used to create a file on the file system, but also to retrieve the log messages again using the IsolatedStorageLogEntry Repository.

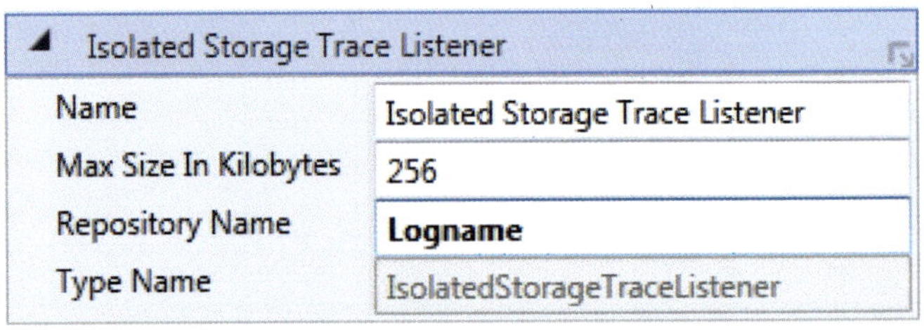

5. Configure the All Events Trace Source to use the Isolated Storage trace listener.

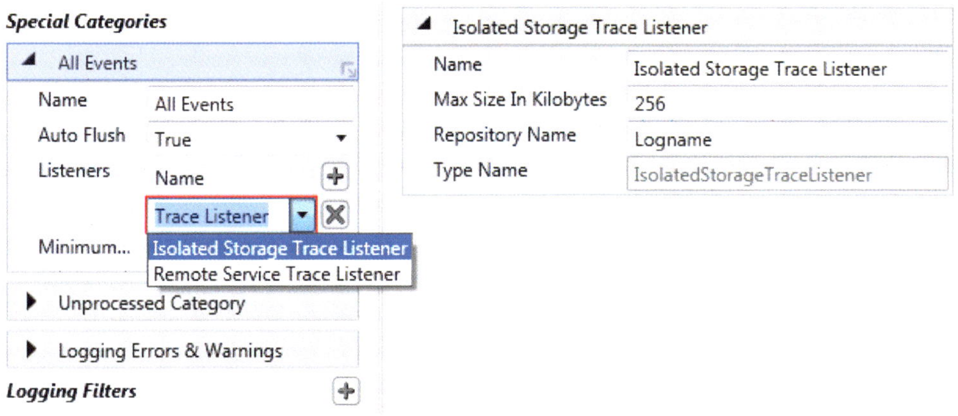

6. Select **Wizards -> Export to XAML**. Browse to your Silverlight project and save it with the name Configuration.xaml.

7. Add the Configuration.xaml file to your Silverlight application project and set the build action to **Resource** or **Page** in the file properties.

Reading the Log Messages from Isolated Storage

You can use the Isolated Storage Log Entry Repository to retrieve log messages that were written to isolated storage. You will need to provide the name of the repository that was specified in configuration.

You can use the following code example to read all log messages that are present in isolated storage:

```
C#
public IEnumerable<LogEntry> GetLogMessages()
{
   // Get the Isolated Storage Log Repository. The Name used
   // here should correspond to the repository name specified
   // in configuration
   IsolatedStorageLogEntryRepository repository =
      EnterpriseLibraryContainer.Current.GetInstance<
      IsolatedStorageLogEntryRepository>("Logname");
   return repository.RetrieveEntries();
}
```

Using the Notification Trace Listener

In this section, we'll discuss how you can use the Notification trace listener so you can respond to an event every time a log message is written.

Configure the Notification Trace Listener

1. Open the Enterprise Library configuration console.
2. Select **Blocks -> Add Logging Settings for Silverlight**

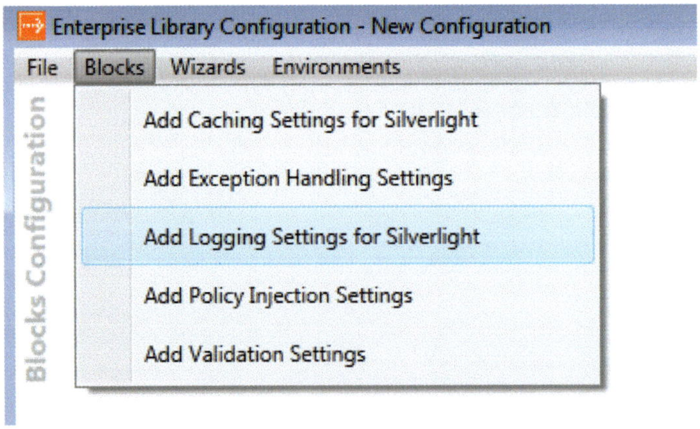

3. Add the Notification trace listener.

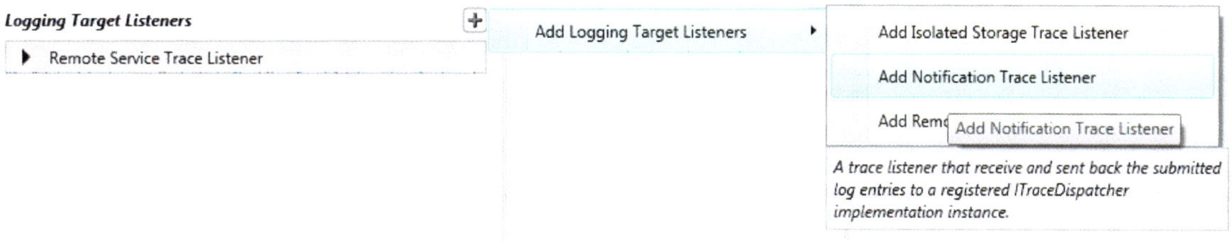

4. Configure the All Events Special Trace Source to use the Notification trace listener.

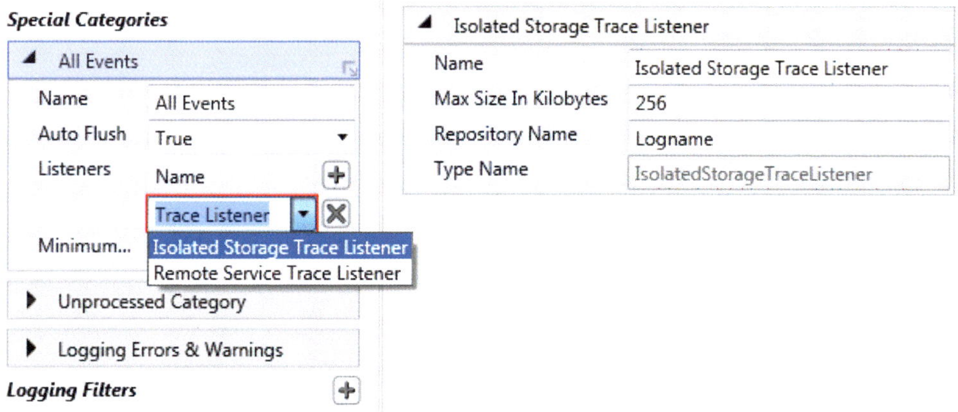

5. Select **Wizards -> Export to XAML**. Browse to your Silverlight project and save it with the name "Configuration. xaml."

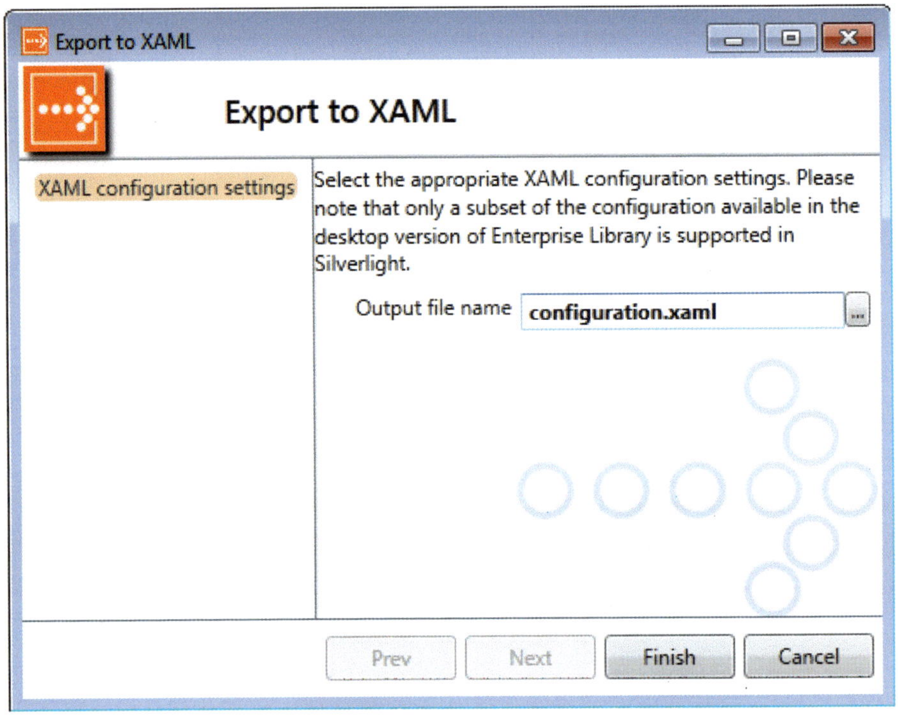

6. Add the configuration.xaml file to your Silverlight application project and set the build action to **Resource** or **Page** in the file properties.

Respond to the Notification Event

To respond to the event when a new log message is written, you will need to get an instance of the ITraceDispatcher and register an event handler to the **TraceReceived** event. The following example shows how this works:

```
C#
public void StartMonitoringForNewLogMessages()
{
   var dispatcher = EnterpriseLibraryContainer.Current.
      GetInstance<ITraceDispatcher>();
```

```
    dispatcher.TraceReceived += dispatcher_TraceReceived;
}
void dispatcher_TraceReceived(object sender,
    TraceReceivedEventArgs e)
{
    // A log entry was received. Handle it as appropriate:
    LogEntry logEntry = e.Data as LogEntry;
}
```

Changing the Logging Levels in the Application

Using the Change API, you can make run-time changes to the Logging Block. To make any changes, you will need to call the **GetUpdate Context** method on the **LogWriter** class to get an **ILogWriter UpdateContext** instance. You can use this instance to change the logging settings. The changes you make will be activated when you call **ApplyChanges()**.

Turning Logging On Or Off

The following code demonstrates how you can turn logging on, if it was disabled in configuration:

```
C#
public static void TurnLoggingOn()
{
    LogWriter logWriter = EnterpriseLibraryContainer.Current.
        GetInstance<LogWriter>();
    ILogWriterUpdateContext updateContext =
        logWriter.GetUpdate-Context();
    updateContext.IsLoggingEnabled = true;
    updateContext.ApplyChanges();
}
```

Changing the Logging Level for a Trace Source

The following code sample demonstrates how you can increase or decrease the logging level for a particular trace source. For example, if your trace source is configured to log only messages with a severity of Error or greater, then you can use this sample to change that setting to Warning or greater.

```
C#
public static void IncreaseTraceLevel()
{
   LogWriter logWriter = EnterpriseLibraryContainer.Current.
      GetInstance<LogWriter>();
   ILogWriterUpdateContext updateContext =
      logWriter.GetUpdate-Context();
   // This will update the special trace sources
   updateContext.AllEventsCategory.Level =
      SourceLevels.Warning;
   updateContext.ErrorsCategory.Level = SourceLevels.Warning;
   // This will update a custom category, with the name "Custom"
   updateContext.Categories.First(c => c.Name == "Custom").Level
      = SourceLevels.Warning;
   updateContext.ApplyChanges(); ;
}
```

Adding a Trace Listener to a Trace Source

The following code sample demonstrates how you can add a trace listener to a trace source.

```
C#
public static void AddTraceListener()
{
   LogWriter logWriter = EnterpriseLibraryContainer.Current.
      GetInstance<LogWriter>();
   ILogWriterUpdateContext updateContext =
      logWriter.GetUpdate-Context();
   // Add a trace listener with the name "Service"
   updateContext.AllEventsCategory.Listeners.Add("Service");
   // This will update a custom category with the name
   // "Custom" to add the trace listener
   updateContext.Categories.First(c => c.Name == "Custom").
      Listeners.Add("Service");
   updateContext.ApplyChanges();
}
```

Note: *The trace listener you are trying to add must already be used by another trace source. You cannot add new trace listeners or use trace listeners that have been configured but are not used by any trace sources.*

Changing the Remote Service Trace Listener

The following code sample demonstrates how you can make run-time changes to the Remote Service Trace Listener. In this sample, we're enabling buffering and increasing the size of the isolated storage buffer to 512 Kb:

C#

```
public static void ChangeRemoteServiceTraceListener()
{
   LogWriter logWriter = EnterpriseLibraryContainer.
      Current.GetInstance<LogWriter>();
   ILogWriterUpdateContext updateContext =
      logWriter.GetUpdate-Context();
   // Get the update context for the trace listener
   var listener = updateContext.Listeners
      .OfType<IRemoteServiceTraceListenerUpdateContext>()
      FirstOrDefault();
   if (null != listener)
   {
      // Set the size of the isolated storage buffer
      // and enable buffering
      listener.IsolatedStorageBufferMaxSizeInKilobytes = 512;
      listener.SendImmediately = false;
      updateContext.ApplyChanges();
   }
}
```

Changing the Isolated Storage Trace Listener

The following code sample demonstrates how you can change the size of the Isolated Storage Log at run time. You should test if the repository is actually available. If there is more than one instance of your application running, then only a single instance can change the repository:

```
C#
public static void ChangeIsolatedStorageTraceListener()
{
    LogWriter logWriter = EnterpriseLibraryContainer.Current.
        GetInstance<LogWriter>();
    ILogWriterUpdateContext updateContext =
        logWriter.GetUpdate-Context();
    // Get the update context for trace listener
    var listener = updateContext.Listeners.First(l => l.Name ==
        "Isolated Storage Trace Listener")
        as IIsolatedStorageTraceListenerUpdateContext;
    // You should test if the repository is available.
    // If you have multiple instances of your application
    // running at the same time, then only a single
    // instance can update the size of the repository
    if (listener.IsRepositoryAvailable)
    {
        // Change the size of the isolated storage trace
        // listener to 5 kb
        listener.MaxSizeInKilobytes = 512;
    }
    updateContext.ApplyChanges();
}
```

More Information

In this chapter, we discussed how you can apply the Logging Application Block to a Silverlight application. To learn more about the Logging Application Block, please read the chapter on the **Logging Application Block on MSDN** and in the **Developer's Guide**.

5 Exception Handling

In this chapter, we'll discuss how you can implement exception handling in a Microsoft® Silverlight® browser plug-in application using the Silverlight Exception Handling Application Block. We'll also discuss how you can apply exception handling to asynchronous services. Then we'll discuss how exception handling was applied in the Stock Trader V2 reference implementation.

Exception Handling Using the Exception Handling Application Block

The Exception Handling Application Block allows you to handle in a consistent manner exceptions that might occur in any layer of your application. The following code shows an example of how to use the **ExceptionManager** class to handle an exception:

C#

```
try
{
   // Code that might throw an exception
}
catch(Exception ex)
{
   exceptionManager.HandleException(exception,
   "ServiceLayer");
}
```

In the configuration console, you can then define how you would like to handle such exceptions, by specifying policies. This example shows how you could handle unhanded exceptions, in this case by logging the exception and displaying a friendly exception message:

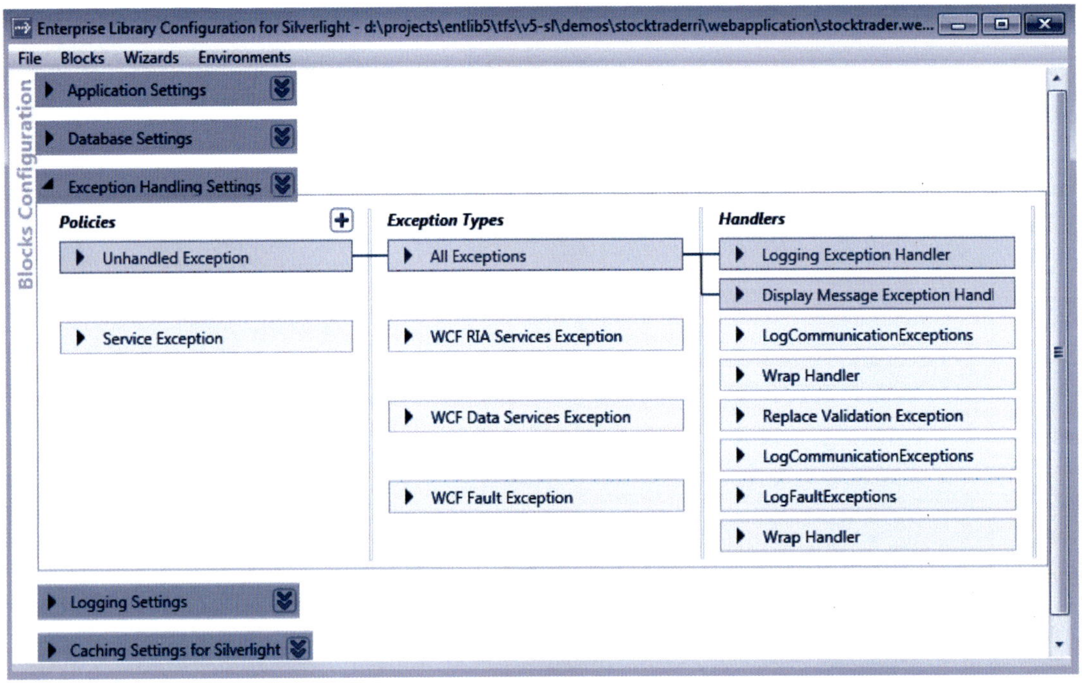

Exception Handling in a Silverlight Application

Exception handling in a Silverlight application is very similar to exception handling in any Microsoft .NET Framework application. The **general guidance** for exception handling in a .NET Framework application also applies to Silverlight applications. There are Silverlight-specific aspects that you should keep in mind, which we will discuss next.

Global Exception Handler

Silverlight provides a convenient place for global exception handling. The **Application** object, located in App.xaml, raises an event if an unhandled exception occurs.

Unlike a regular Microsoft Windows® application, a Silverlight application will not automatically shut down when an unhandled exception occurs. By default, when you create a new Silverlight application, it will automatically add an unhandled exception handler for you that will report an unhandled exception to the browser. In most cases, the application will then keep running, without exiting.

This behavior is similar to that of a regular web application when an exception occurs in JavaScript. The advantage of this behavior is that the application can continue if the exception was not very serious. However, allowing the user to continue after an unhandled exception will leave the application in an unknown state. This can result in other problems such as data loss or data corruption. You should consider what the consequences would be for your application if the user were to be allowed to continue after an unhandled exception.

During our usability tests we found that our users like to be informed if something goes wrong. They want to know what has happened, but also what they should do to solve it. In the Stock Trader V2 reference implementation, if a call to a web service fails, we had to explicitly handle that action in our View Models. Depending on the situation, we either needed to show a message and automatically retry, or show a message that something went wrong and allow the user to retry or cancel the action.

Handling Exceptions in Asynchronous Web Services

Because Silverlight only supports asynchronous web services, you'll have to take extra care when implementing exception handling for your services. It's usually a good idea to implement exception handling at the service boundaries, because it is likely that some service calls will fail. A small hiccup in the user's network connection can be enough for a web service call to fail. In those cases, you should notify the user and allow him to retry.

When you're calling a synchronous method or service and an exception occurs, that exception will travel back up the call stack until you decide to handle it using a try-catch statement.

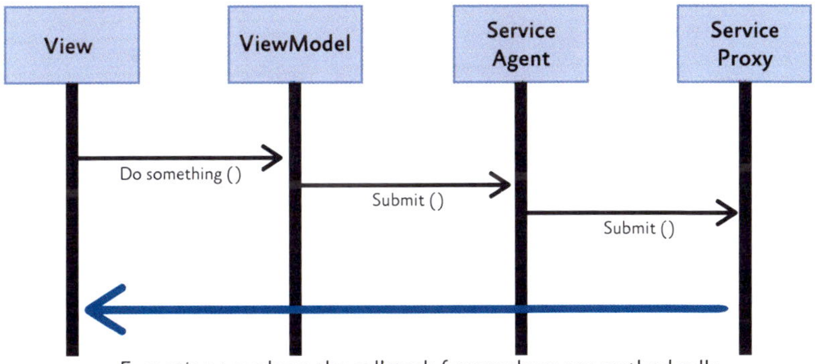

Exception travels up the callstack for synchronous method calls

However, if an exception occurs while calling an asynchronous web service, then it is up to the proxy to raise an event or execute a callback to indicate that an exception has occurred. If you don't handle that exception, it will not bubble back up the call stack, but will immediately pass to the unhandled exception handler. So if you want to notify the view model or the view that a problem occurred, you will need to explicitly pass the error information from the service agent to the ViewModel and eventually to the view to show an error message.

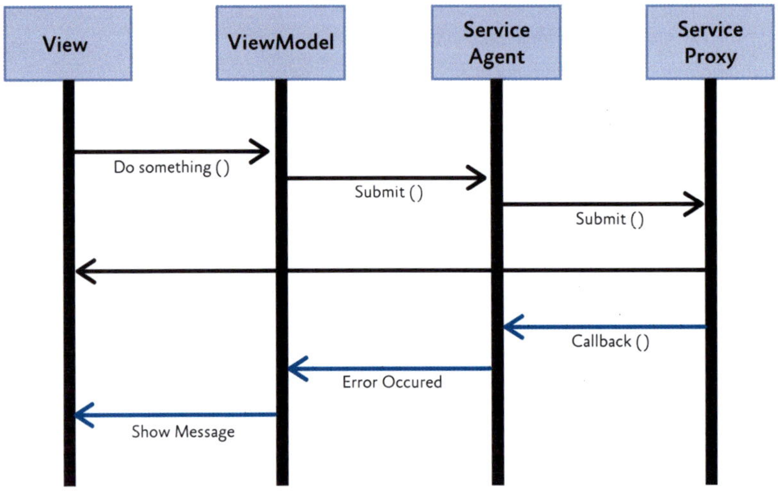

If you don't handle the exception when it occurs, it will not travel up the call stack automatically, but instead will trigger the global unhandled exception handler. In the global exception handler, you are able to show a generic error message, but it is usually not possible to change the state of the UI to allow the user to retry. This chapter will give you a short introduction to the Enterprise Library Exception Handling Application Block. Since the Exception Handling block has been completely ported to Silverlight, with the exception of the Windows Communication Foundation (WCF) fault contract exception handler, this chapter will provide only a short introduction. If you'd like more information about the Exception Handling block, please read the chapter on the **Exception Handling Application Block** on MSDN®.

Introduction

The Silverlight Exception Handling Application Block helps developers create a consistent strategy for processing exceptions. When an exception occurs, it's very common to log the exception, wrap the exception in a different exception, or replace the exception entirely. Instead of duplicating this code throughout your application, the Exception Handling block allows you to define policies to determine how exceptions should be handled.

The following diagram shows how the Silverlight Exception Handling Application Block works:

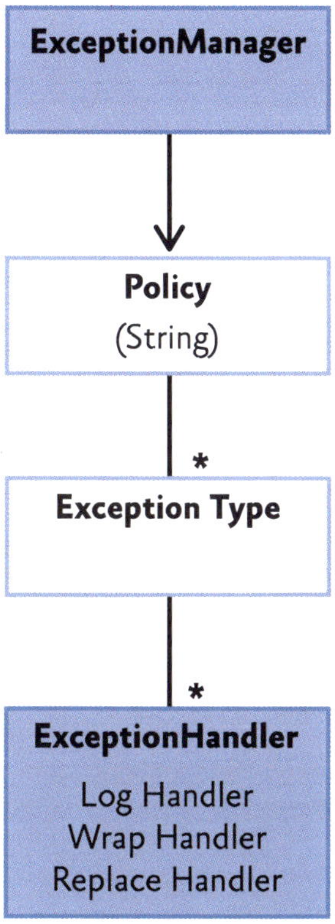

The Exception Manager can be configured with one or more policies. A policy, which is defined by its name, can target one or more exception types. Each exception type can have one or more exception handlers, such as the log handler or the wrap handler.

For example, you could create a policy called Unhandled Exceptions, that targets all exceptions and that will first log the exception using the Log Handler and then display a user-friendly error message using a custom exception handler. You might also create a policy for your service agents, called Service Agents. This policy would target specific exceptions that might occur when accessing web services. When such an exception occurs, it might log the error message and then wrap the exception in a more generic **ServiceException**.

Exception Handling Application Block

Features

The Silverlight Exception Handling Application Block supports the following scenarios.

Handling Exceptions

By using the **Handle Exception** method on the Exception Manager class, you can tell the Exception Handling block to handle a specific exception. To avoid having to create try-catch statements in which to place the **HandleException** method, you can also use the **Process** method on the **ExceptionManager** class. The **Process** method of the **ExceptionManager** class executes the code you specify in the delegate (method). If an exception occurs while this code is executing, the **Process** method sends the exception to the Exception Handling Application Block to be handled using the policy you specify in the parameters of the **Process** method.

For more information, please read the chapter on Sending an Exception to the Exception Handling Application Block on MSDN®.

Defining Exception Handling Policies

There are two ways you can define exception handling policies. You can either define them in a XAML configuration file or use the fluent configuration API to define your exception handling policies.

For more about configuring the Exception Handling Application Block, please read about Entering Configuration Information on MSDN®.

Exception Handlers

- The Silverlight Exception Handling Application Block ships with the following exception handlers
- **Wrap handler**. This wraps one exception around another.
- **Replace handler**. This replaces one exception with another.
- **Logging handler**. This formats exception information, such as the message and the stack trace. Then the logging handler passes this information to the Enterprise Library Logging Application Block so that it can be published.

Exception Handling in the Stock Trader V2 Reference Implementation

In this chapter, we'll discuss how we applied exception handling in the Stock Trader V2 reference implementation. Exception handling is a true cross cutting concern, so it was applied to all modules. The News Module also applies exception handling using interception, but this is described in the chapter on Interception in the Stock Trader: The News Module.

Goals for Exception Handling in the Stock Trader V2

When we started to work on the reference implementation, we had the following goals:

- **Dealing with unhandled exceptions:** If an unhandled exception occurs, we want to display a user-friendly error and exit the application.
- **Unified service exception handling mechanism:** The Stock Trader V2 uses several web service technologies, such as plain WCF services, WCF RIA Services and WCF Data Services. Each of these technologies has its own way of raising exceptions on a callback. These implementation details should be hidden by the service agent classes, so the service agents should provide a uniform mechanism for notifying the calling classes of exceptions.
- **Retrying after service exceptions:** If a web service call fails for some reason, then the user should have the option to retry. If the web service call is the result of a button click, then the user should be able to click the button again. However, if a screen needs to load some data before it can be displayed, and the web service call to retrieve that data fails, then the UI should display a screen with a friendly error message and a retry button.

Unhandled Exceptions

Catching unhandled exceptions is really easy in Silverlight. By default, the App.xaml.cs file, which is located in the root of the StockTraderRI. Silverlight project, contains an event handler called **Application_UnhandledException**. This is an easy place to perform centralized exception handling.

The following code comes from the **Application_Unhandled Exception** event in the App.xaml.cs file, which demonstrates how unhandled exceptions are caught and then handled by the Exception Manager:

```
C#
if (!enterpriseLibraryConfigured)
{
   // If enterprise library is not configured. Just show
   // an error message and exit (The error window will exit the
   // application when the user closes the error message)
   App.Current.RootVisual = new UserControl();
   ErrorWindow.CreateNew("An unknown problem has occurred while
      starting the application", e.ExceptionObject,
      Guid.Empty, true);
   return;
}
// Call the Unhandled exception policy. This should:
// 1. Log the error
// 2. Show a friendly error message
// 3. Exit the application (The DisplayMessageExceptionHandler
// will do this)
var exceptionManager = EnterpriseLibraryContainer.Current
   .GetInstance<ExceptionManager>();
exceptionManager.HandleException(e.ExceptionObject,
   "Unhandled Exception");
// set the exception to handled
e.Handled = true;
```

First of all, if an exception has occurred while Enterprise Library was being configured, then we can't rely on Enterprise Library to handle the exception properly. In that case, we're just showing an error message.

The following diagram shows how the exceptions get handled by the Exception Manager:

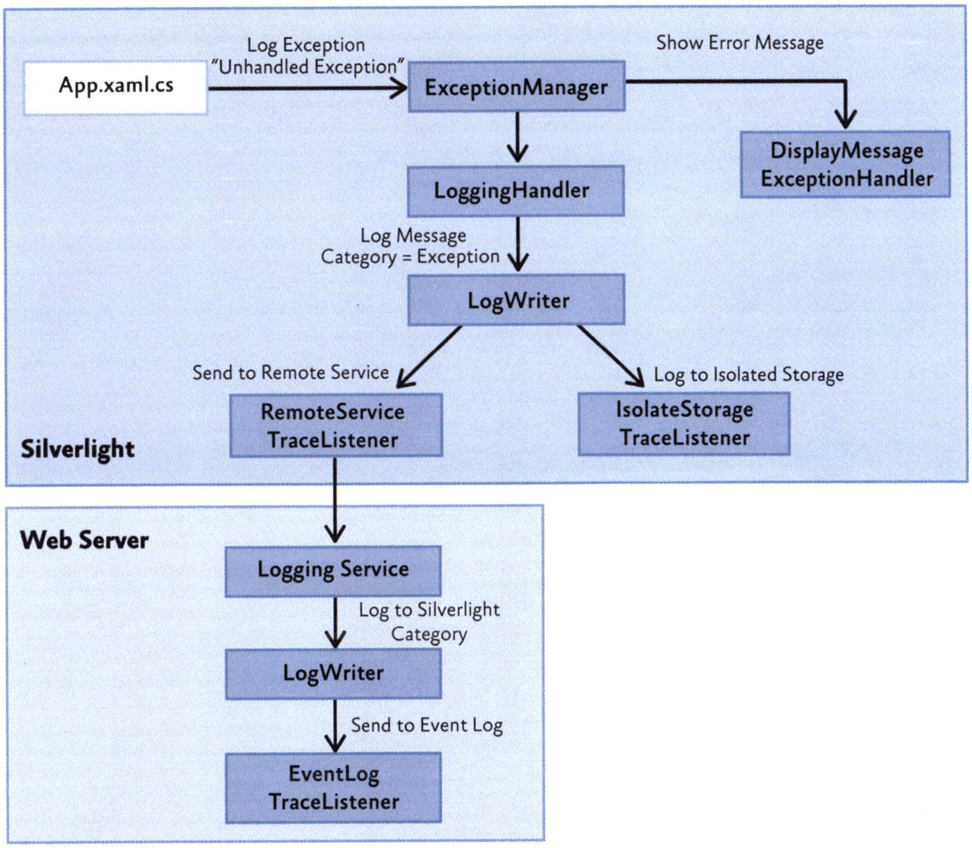

The unhandled exceptions are handled using the Unhandled Exception policy. This policy is configured to:

- **Log the exception**
  The Exception Handling block will forward the exception to the Logging Application Block, to a category called Exception, with severity of Critical. In the configuration for the logging block, this category is configured to be logged to the Remote Service trace listener, to the log file in isolated storage and to the developer trace window using the Notification trace listener. The Remote Service trace listener sends it back to the server. There, it will be handled by the Enterprise Library 5.0 Logging Application Block. This block is configured so that all log messages that are sent to the server are written to the Windows Event Log.

- **Show a friendly error message**
  A custom exception handler, called **DisplayMessageException Handler**, was written to display a friendly error message to the user. In order to use this class through configuration, we also needed to create a corresponding display error message exception handler data class. The **DisplayMessageExceptionHandler** can also be configured to exit the application after showing the technical error message.

Handling Exceptions When Calling Services

In Stock Trader V2, all interactions with web services are encapsulated within special service agent classes. These service agents abstract the technical implementation details for accessing the web services from the rest of the application. They are also very well suited to applying exception handling to web services.

Not all of the techniques described in this chapter use the Exception Handling block. They are more of a general description of how you can apply exception handling with asynchronous services.

Returning the Results from a Service

The service agents all use a callback action to notify the caller that the service has completed. This callback is also responsible for handing the result of the service callback to the caller. For example, the order service agent has a method to retrieve a list of orders. If this method executes successfully, then the callback should pass the list of orders to the caller. However, if there was an exception while accessing the web service, then the callback should return the exception that occurred to the calling method.

To make this process more manageable, two classes have been introduced. The **AsyncServiceRequest** and **ServiceResult** classes. The **AsyncServiceRequest** encapsulates the callback to notify the user interface that the download request has completed. It has several helper methods that make it easier for the service agent to either return the requested data or return an error.

```
C#
public class AsyncServiceRequest<T> : IAsyncServiceRequest
  where T : ServiceResult, new()
{
  // Creates the Service Request class, which wraps the
```

```
    // callback action
    public AsyncServiceRequest(Action<T> callback, object
        user-State = null){}
    // Typically, this class is passed as userstate to an async
    // web service call. This method helps to restore it from
    // the user state.
    public static AsyncServiceRequest<T>
    RestoreFromUserState(object userData){}
    // It's common to have to provide additional user state.
    public object UserState { get; set; }
    // Helps the service agent to return a list of validation
    // errors to the UI.
    public ServiceResult ReturnValidationErrors(
        ValidationErrorException validationErrorException) {}
    // Helps the service agent to return a regular result
    // to the UI.
    public ServiceResult ReturnResult(T result){}
    // Helps the service agent to return an error to the UI.
    public ServiceResult ReturnError(Exception exception){}
    // Helps the service agent return an empty result to the UI
    public ServiceResult ReturnEmpty() {}
    // Helps to implicitly convert an action to an
    // AsyncService-Request
    // to reduce the need to create one explicitly.
    public static implicit operator AsyncServiceRequest<T>(
        Action<T> requestCompleteCallback){}
    // Event that gets fired before the callback is executed
    // to allow the callback to be intercepted and altered.
    public event EventHandler
        <AsyncServiceRequestOnCallbackEventArgs>
    OnCallback = delegate { };
}
```

The **ServiceResult** class encapsulates the result of a service call. This can either be the data that was actually requested, a set of validation results, or a technical exception that has occurred. There is also a generic variant of the ServiceResult, which makes it easier to program in a typed way, and a **ServiceListResult** class, which makes it easier to return a list of items.

All service agents use the **AsyncServiceRequest** class to start an asynchronous download and the ServiceResult to notify the caller of the results.

This is the signature of the **ServiceResult** and **ServiceListResult** class:

C#

```
[DataContract]
public class ServiceResult<T>
{
   public IEnumerable<ValidationResult> ValidationResults
      { get; private set; }
   public Exception Error { get; set; }
   public bool HasError { get; set; }
   public bool IsHandled { get; set; }
   public object UserState { get; set; }
   [DataMember]
   public T Result { get; set; }
}
public class ServiceListResult<T> : ServiceResult<IEnumerable<T>>
{}
```

You can see that the **ServiceResult** has properties for passing the result, but also properties for conveying exceptions and server-side validation errors.

In the **OrderServiceAgent**, the method to retrieve the list of orders has the following signature:

C#

```
public void LoadOrders(AsyncServiceRequest<
   ServiceListResult<Order>>
   callbackToUpdateUI, object userState)
```

You can see that the **LoadOrders** method receives an object of the type **AsyncServiceRequest** of **ServiceListResult** of **Order**. This means that the callback will return either a list of orders, or an exception.

In the **OrderListViewModel**, this is how the service list result is used:

```
C#
public void StartLoadingOrders()
{
   // Start to load the orders asynchronously, and call the
   // OnOrdersLoaded Method when done.
   Action<ServiceListResult<Order>> asyncServiceRequest =
      OnOrdersLoaded;
   ordersServiceAgent.LoadOrders(asyncServiceRequest, null);
}
public void OnOrdersLoaded(ServiceListResult<Order> result)
{
   // The order has been loaded.
   if (result.HasError)
   {
      // There were errors. Display a friendly error message.
      // Mark the error as handled. If you don't mark the error
      // as handled it will get rethrown.
      result.IsHandled = true;
   }
   else
   {
      // Show the Orders in the user interface.
   }
}
```

You can see that the **ViewModel** has to explicitly set the **IsHandled** property to true, when it is finished with handling the exception. If not, it will be rethrown and trigger the global unhandled exception handler.

For some service agent calls, not actually handling the exception is the preferred way of working. If it is ok to display the generic "Something is wrong" friendly error message, then you can simply not handle it in the ViewModel. However, in the Stock Trader V2, there are many places where we explicitly want to show a custom message or custom user interface when a service call fails. This cannot be done using an exception handling policy, so it needs to be handled in the ViewModel itself.

Correlating the Start and End of an Asynchronous Operation

The asynchronous methods provided by all the service agents allow you to pass in a callback, wrapped inside an AsyncServiceRequest, that should be executed when the asynchronous action is complete. Here, it is important to correlate the start of the asynchronous action with the completion. The following code sample, from the account holder service agent, demonstrates how you can correlate the start and the end of the callback.

```
C#
public void GetAccountHolder(
   AsyncServiceRequest<AsyncServiceRequest<AccountHolder>>
   asyncServiceRequest,
   object userState)
{
   query = context.AccountHolderSet.Expand("BankAccounts");
   // Begin the asynchronous action that will load the
   // Account Holder; the AsyncServiceRequest is then passed
   // along as the userstate, which can be restored when the
   // async action completes.
   query.BeginExecute(OnGetAccountHolderComplete,
      asyncService-Request);
}
```

The different technologies for executing web services, such as WCF, WCF Data Services and WCF RIA Services, all have different techniques for starting an asynchronous action and notifying that an asynchronous action has completed. However, all of these techniques allow you to pass an object as user state. This user state is then passed back when the service request completes.

This code sample shows how you can restore the Async Service Request when the service returns and executes the correct callback:

```
C#
private void OnGetAccountHolderComplete(IAsyncResult ar)
{
   // The ar.AsyncState is an object. Cast it to an
   // AsyncServiceRequest class
   var asyncServiceRequest = AsyncServiceRequest<ServiceResult<A
      ccountHolder>>.RestoreFromUserState(ar.AsyncState);
   List<AccountHolder> resultList;
   try
   {
      // If a problem occurred while executing the query,
      // then WCF Data Services will throw that exception when
```

```
        // you call the 'EndExecute' method. This
        // exception will be handled by the catch block.
        resultList = query.EndExecute(ar).ToList();
    }
    catch (Exception ex)
    {
        // Handle the exception using the exception manager, and
        // use the callback to report the error to the UI.
        exceptionManager.HandleAsyncWcfServiceError(
            ex, asyncServiceRequest, ExceptionHandlingPolicies
            .ServiceLayer);
        return;
    }
    // Try to get the account holder. Create it if it does
    // not exist.
    var result = resultList.FirstOrDefault() ??
        CreateNewAccount-Holder();
    // Execute the callback to hand the result back to the UI.
    asyncServiceRequest.ReturnResult(new ServiceResult<
        AccountHolder>(result)
    {
        // Pass the user state that was passed with the
        // GetAccountHolder method back to the calling method.
        UserState = asyncServiceRequest.UserState
    });
}
```

The **IAsyncResult** that's passed by WCF Data Services, also contains the user state that was used when starting the request as an object. This user state will contain the Async Service Request instance that was created by the **GetAccountHolder** method.

The Async Service Request contains both the callback method that should be used to update the UI, and the user state that was passed to the **GetAccountHolder** method.

Handling Exceptions in the Asynchronous Service Agent

When an exception occurs in a synchronous method that you wish to handle, you can easily add a try-catch block and call the **Handle Exception** method of the **ExceptionManager** class. With the asynchronous actions in the service agents, this is not as simple. You will always need to notify the UI that the asynchronous action has completed, but the details depend on the actual settings in your exception handling policy.

Stock Trader V2 uses the following helper method to handle asynchronous WCF Service exceptions:

```
C#
public static void HandleAsyncWcfServiceError(
    this ExceptionManager exceptionManager,
    Exception exception,
    IAsyncServiceRequest asyncServiceRequest,
    string policy)
{
    // Handle the exception using the supplied exception message.
    Exception errorToThrow;
    var rethrow = exceptionManager.HandleException(exception,
    policy, out errorToThrow);
    if (rethrow)
    {
        // The exception is not swallowed. That means the UI
        // needs to be notified of the exception.
        errorToThrow = errorToThrow ?? exception;
        if (errorToThrow is ValidationErrorException)
        {
            asyncServiceRequest.ReturnValidationErrors(
            errorToThrow as ValidationErrorException);
        }
        // Notify the UI of the problem.
        var result = asyncServiceRequest.
            ReturnError(errorToThrow);
        if (!result.IsHandled)
        {
            // The UI didn't handle the exception. That means we
            // need to rethrow the exception this will trigger the
            // global exception handling logic in App.xaml.cs.
            throw errorToThrow;
        }
    }
    else
    {
        // The exception is swallowed by the exception handling
        // block. However, you will still need to notify
        // the UI that the exception was handled.
        asyncServiceRequest.ReturnEmpty();
    }
}
```

The exception needs to be handled by the exception handling block. In our service agents, we have configured the exception handler to log the error and replace it with a more generic **ServiceUnavailable Exception**. However, this behavior can easily be changed by changing the configuration.

If the post handling action is set to **None**, you are telling the exception handling block to effectively swallow the exception. Since the service threw an exception, there is no result to pass back to the UI. However, the UI still needs to be notified that the asynchronous action has completed without a result.

If you have set the post handling action either to **NotifyRethrow** or **ThrowNewException**, then you should notify the UI that an exception has occurred. The **ServiceResult** class also allows you to pass an exception back to the UI. In that case, the **HasError** property is set to true.

If the user interface handles the error, then no further action is needed. However, if the UI does not handle the exception, then the exception should be rethrown. Throwing the exception again will trigger the global **Unhandled Exception** event.

Except for the unhandled exception handler, it is somewhat unusual to swallow exceptions using **None** as the post-handling action. Not returning a result is usually not something developers expect from a web service call and if they forget to test the result value for nulls, the outcome could be other errors.

Displaying Error Messages

The Stock Trader V2 also uses a custom exception handler to display exception messages. The advantage is that you can display the exception handling instance ID to the end user. This instance ID is also logged, to help the support organization find the particular error if the user reports it.

The following code shows how this custom exception handler displays an error window.

C#
```
public class DisplayErrorMessageExceptionHandler :
   IExceptionHandler
{
   private readonly ErrorMessageType errorMessageType;
   private IThreadingSupport threadingSupport = new Threading-
      Support();
   public DisplayErrorMessageExceptionHandler(ErrorMessageType
      errorMessageType)
   {
      this.errorMessageType = errorMessageType;
   }
   public Exception HandleException(Exception exception, Guid
      handlingInstanceId)
   {
```

```
        threadingSupport.ExecuteOnUIThread(() => ErrorWindow.
          CreateNew(exception, handlingInstanceId,
          errorMessageType ==
          ErrorMessageType.TechnicalErrorMessage));
        return exception;
    }
}
public class DisplayErrorMessageExceptionHandlerData :
    Exception-HandlerData
{
    /// <summary>
    /// Get the set of <see cref="TypeRegistration"/> objects
    /// needed to register the call handler represented by this
    /// config element and its associated objects. </summary>
    /// <param name="nameSuffix">A suffix for the names in the
    /// generated type registration objects.</param>
    /// <returns>The set of <see cref="TypeRegistration"/>
    /// objects.</returns>
    public override IEnumerable<TypeRegistration>
    GetRegistrations(string nameSuffix)
    {
        yield return
        new TypeRegistration<IExceptionHandler>(
          () =>
        new DisplayErrorMessageExceptionHandler(Error
          MessageType))
        {
          Name = BuildName(nameSuffix),
            Lifetime = TypeRegistrationLifetime.Transient
        };
    }
    public ErrorMessageType ErrorMessageType { get; set; }
}
```

More Information

In this chapter, we discussed how you can use the Exception Handling Application Block in a Silverlight application.

To read more about exception handling using the Exception Handling Application Block, please visit the **online documentation** on MSDN® and the **Developer's Guide**.

6 Validation

Validation is a very important aspect of any line of business (LOB) application. A good validation implementation helps prevent data corruption and enforces compliance with business rules.

Microsoft® Silverlight® allows you to build applications that provide a rich user experience, and part of a rich experience is good validation. When a validation error occurs, the user should get clear and obvious feedback as quickly as possible so he is not left wondering what is happening. Simple validation checks might be performed as the user is typing, or more complex validation rules might be run in the background while the user is still performing some task. Validation error messages should provide a clear description of the cause of the error, with an indication of how the user can solve the problem.

In this chapter, we'll examine the most common validation scenarios for Silverlight LOB applications and how the Validation Application Block from the Enterprise Library 5.0 Silverlight Integration Pack can help you implement those scenarios in your application. We'll also discuss how the Validation Application Block can be applied to a Windows® Communication Foundation (WCF) Data Services application and how you can use the Validation Application Block to extend the validation support provided by WCF RIA Services.

Validation in a Silverlight Application

A Silverlight LOB application typically consists of a Silverlight client that communicates with web services to read or modify data in data sources such as databases. As the data flows through the user interface all the way to the data source, there are several places to handle validation. Validation errors that occur in this process will typically flow back to the user interface (UI):

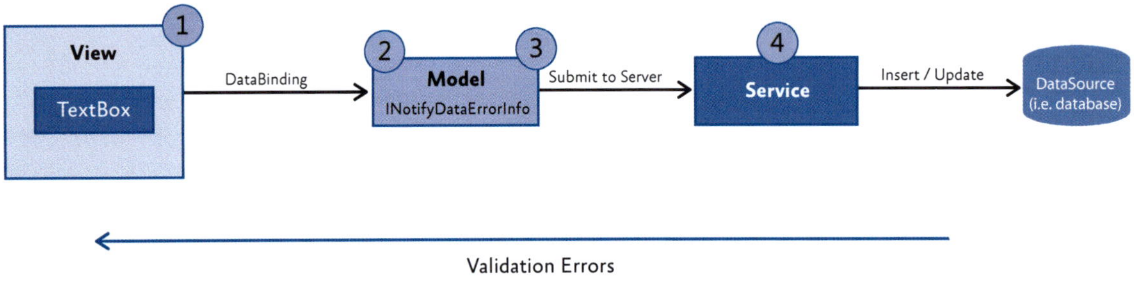

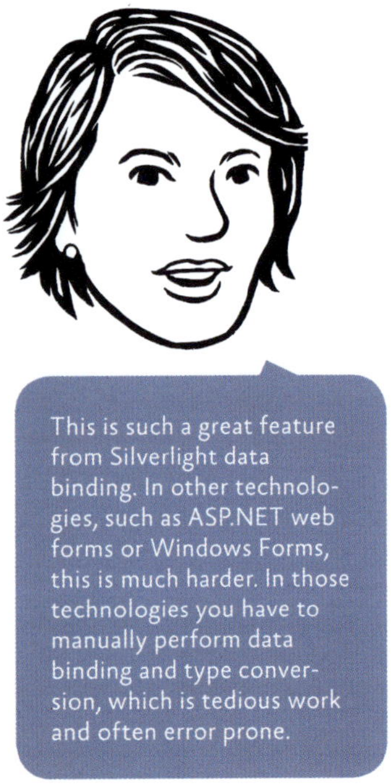

1. The user enters data in a control, for example a **TextBox**, that is bound to a property on your model. In this context, a model can refer to a View Model from the Model View ViewModel (MVVM) pattern, but also to an entity class generated by WCF RIA Services or a custom business entity. When the user leaves the **TextBox**, the Silverlight data binding engine will attempt to convert the value in the **TextBox** to the type of the property and update the model. If the value in the **TextBox** cannot be converted to the type of bound property, then the data binding engine will throw an exception. You can then use an error style to display an error message.
2. A rudimentary set of validation rules can be called as soon as the model is updated. For example, you can check if the value falls within certain bounds or fits a specific pattern such as an email address or telephone number. When such a validation error occurs, the model will inform the UI that a validation error has occurred. This can be done by raising an exception that will be handled by the data binding engine or by implementing **IDataErrorInfo**. However, a more flexible approach is to implement **INotifyDataErrorInfo** on the model and raise the **ErrorsChanged** event.
3. When the user is finished entering all information and is ready to submit, some more complex validation rules can be executed. Typically, these are validation rules that span multiple properties or check against reference data. Again, if a validation error has occurred, the UI has to be notified. Throwing an exception in this situation is not an option because the data binding engine cannot catch this exception in that case. For these validation rules, you should implement **INotifyDataErrorInfo** on your model and raise the **ErrorsChanged** event.

4. When there are no validation errors, the data can be submitted to the server. Since validation rules on the client can easily be circumvented, you must run all critical validation rules again on the server. It is also possible that there are additional validation rules that should only be run on the server. Because the web server is typically closer to the data sources (such as the database), you'll be able to do other types of validation checks, such as checking for referential integrity within a transaction. Any validation errors that occur here will also have to be sent back to the client and displayed in the UI. Again, **INotifyDataErrorInfo** can help in this regard.

The Validation Application Block

The Enterprise Library Validation Application Block provides useful features that allow developers to implement structured and easy-to-maintain validation scenarios in their applications.

The Validation Application Block has been completely ported from Enterprise Library 5.0 to the Enterprise Library 5.0 Silverlight Integration Pack.

To learn more about the Validation Application Block, you can read more in the **Developer's Guide** or in the **online documentation** on MSDN®.

Supported Scenarios

The following scenarios are supported by the Validation Application Block from the Enterprise Library 5.0 Silverlight Integration Pack:

- **Execute validation rules and gather results.** Using the **Validator** class, you can validate an object or graph of objects and gather the validation results.
- **Annotating your business entities with validation attributes.** You can define validation rules by annotating your business entities with validation attributes, such as the **StringLength ValidatorAttribute** and the **RangeValidatorAttribute**.

> I prefer to define validation rules by annotating my business entities with validation attributes. This makes the validation rules easy to comprehend and maintain because the code is declarative and describes itself. And also, it's really easy to test your validation rules with unit tests. Since most validation rules don't change often, I can define most of my rules in code and only define them in configuration if they are likely to change often.

- **Data annotations compatibility.** The validation attributes from Enterprise Library are compatible with WCF RIA Services. There are some limitations, which are described in the section on Using the Validation Application Block in a WCF RIA Services application.
- **Defining validation rules in configuration.** Using the Validation Application Block, you can also define validation rules for your business entities in a configuration file written in XAML.
- **Conditional validation using rule sets.** The Validation Application Block supports the concept of rule sets. A rule set is a name given to a group of validation rules. You can use rule sets to run a specific set of validation rules in specific situations.
- **Implementing self-validation.** Some validation rules are hard to express using attributes or configuration. You can also define validation methods on your business entities that will be executed when your objects are validated.
- **Defining validation attributes in metadata.** Silverlight doesn't support the **MetadataTypeAttribute**. In the .NET Framework, this attribute is used to define metadata classes with validation attributes for your generated business entities. The Validation Application Block provides an implementation of the **Meta dataTypeAttribute** for Silverlight.
- **IDataErrorInfo support.** You can use the **DataErrorInfoHelper** to implement **IDataErrorInfo** on your classes. There is no built-in support for **INotifyDataErrorInfo** or **INotifyProperty Info** in the Validation Application Block, but the Stock Trader V2 reference implementation contains several classes to help with implementing these interfaces.

Defining Validation Rules Using Attributes

The most common way to add (relatively simple) validation rules to your business entity is by adding validation attributes to its properties.

The following code snippet shows two examples of adding validation rules through attributes:

```
C#
[RelativeDateTimeValidator(-1000, DateTimeUnit.Year, -18, Date-
   TimeUnit.Year, MessageTemplate = "You must be 18 years or
   Older to use this application.")]
public DateTime DateOfBirth{ get; set }
[NotNullValidator]
```

```
[StringLengthValidator(1, RangeBoundaryType.Inclusive, 0, Range-
   BoundaryType.Ignore, MessageTemplate = "The lastname is
   required")]
public string LastName {get;set;}
```

As you can see, it's possible to combine validation rules by combining several validators together.

Supported Validators

All of the validators that are supported in the desktop version of Enterprise Library are also available in the Enterprise Library 5.0 Silverlight Integration Pack.

Attribute	Description
And Composite Validator	This validator creates a composite validator. Validation requires that all the validators that make up the composite validator be true. For example, you can use the And composite validator to require that both the not null validator and the date time range validator be true.
Contains Characters Validator	This validator checks that the property contains any or all of the characters that are specified by the **CharacterSet** property.
Date Time Range Validator	This validator checks that a **DateTime** object falls within the specified range.
Domain Validator	This validator checks that a value is one of the specified values in a specified set. For example, it can check that a name is "Tom," "Dick," "Harry," or "George" or that an integer is 2, 3, 5, 7, or 11. If the set only contains one value, you can use this validator to check for equality.
Enum Conversion Validator	This validator checks that a string can be converted to a value in a specified enum type. For example, it can check that "Blue" can be converted to a value in the Color enumeration.
Not Null Validator	This validator checks that the value is not **null** (C#) or **Nothing** (Microsoft Visual Basic® development system).
Object Collection Validator	This validator checks that the object is a collection of the specified type and then invokes validation on each element of the collection. Note: This validator will throw an exception if used on a WCF RIA Services entity without specifying a rule set.

Object Validator	This validator causes validation to occur on an object reference. All validators defined for the object's type will be invoked, just as if the **Validation.Validate** method had been called on the object. Note: This validator will throw an exception if used on a WCF RIA Services entity without specifying a rule set.
Or Composite Validator	This validator creates a composite validator. Validation requires that at least one of the validators that make up the composite validator be true. For example, you can use the Or composite validator to require that the Not Null validator or the date time range validator be true.
Property Comparison Validator	This validator compares the value to be checked with the value of a property.
Range Validator	This validator checks that a value falls within a specified range.
Regular Expression Validator	This validator checks that the value matches the pattern specified by a regular expression.
Relative Date Time Validator	This validator checks that the **DateTime** value falls within a specified range using relative times and dates.
String Length Validator	This validator checks that the length of the string is within the specified range.
Type Conversion Validator	This validator checks that a string can be converted to a specific type. For example, the validator can check that "6.32" can be converted to a **Double** type or that "2007-02-09" can be converted to a **DateTime** type.

If you'd like to learn more about applying validation attributes, please refer to 'Using Validation Block Attributes' in the **Enterprise Library Documentation** on MSDN.

Comparing WCF RIA Services' Validation to the Validation Application Block

WCF RIA Services is a technology that integrates the ASP.NET and Silverlight platforms. WCF RIA Services provide a pattern to write application logic that runs on the middle tier and controls access to data for queries, changes, and custom operations. It also provides end-to-end support for common tasks such as data validation, authentication, and roles by integrating with Silverlight components on the client and ASP.NET on the middle tier.

There is quite a bit of overlap in the functionality of WCF RIA Services and the Validation Application Block when it comes to validation. For example, both provide a mechanism to add validation rules to business entities using attributes, both provide an API to trigger validation, and both provide mechanisms to notify the view of validation errors.

This might seem a bit confusing, especially if you're trying to decide which technology to use in your application. While the Validation Application Block can be used standalone, it also works with and extends the validation support that WCF RIA Services provides with some more advanced scenarios. For example, the Validation Application Block also supports validation rules defined in configuration, or conditional validation rules by defining rule sets.

While the Validation Application Block and WCF RIA Services work well together, there are some caveats you'll need to be aware of. These will be discussed in the section called **Using the Validation Application Block in a WCF RIA Services application.**

If you wish to learn more about WCF RIA Services, please refer to the **WCF RIA Services** online documentation on MSDN.

Changes to Enterprise Library 5.0

The entire Validation Application Block has been ported from Enterprise Library 5.0 to the Enterprise Library 5.0 Silverlight Integration Pack. Several changes had to be made to Enterprise Library 5.0 for the desktop, to support:

- **Public properties on validation attributes.** The WCF RIA Services code generator requires that the constructor parameters for all the validation attributes also have corresponding public read-only properties. These have been added.
- **The ToString method implemented on ValidationResult.** The **ValidationResult** class now overrides the **ToString** method, which returns the message of the validation result. This allows you to use data binding to show the validation results in the user interface.
- **The DomainValidator attribute now has additional constructors.** The **DomainValidator** attribute allows you to check a property value against an array of domain values. The WCF RIA Services code generator doesn't support generating attributes with arrays as constructor parameters. To overcome that, additional constructors were added that take up to 10 domain values.

While most validation rules can be executed both on the client and the server, there are some exceptions. Some validation rules can only be executed on the server; for example, checks for referential integrity. But there are also some validation rules which do not need to be executed on the server. For example, in the Stock Trader V2 application, we have implemented some validation rules as warnings. These validation rules are only intended to improve the user experience so they don't need to be executed on the server.

- **The LowerBound and UpperBound properties' types of the RangeValidator and DateTimeRangeValidator types changed to object.** This was required to support the WCF RIA Services code generator.

Performing Cross-Tier Validation Using the Validation Application Block

A Silverlight LOB application typically consists of a Silverlight client that communicates with web services. To improve the user experience, you'll want to run as many validation rules as possible on the client. However, because it's possible to circumvent the client side validation logic, you should also perform the critical validation rules on the server.

Duplicating the validation rules is not a good idea, because it's hard to keep them in sync. It's a better idea to define a validation rule in a single location and if required share it between the client and the server.

The following diagram shows several options for sharing validation rules between the client and the server.

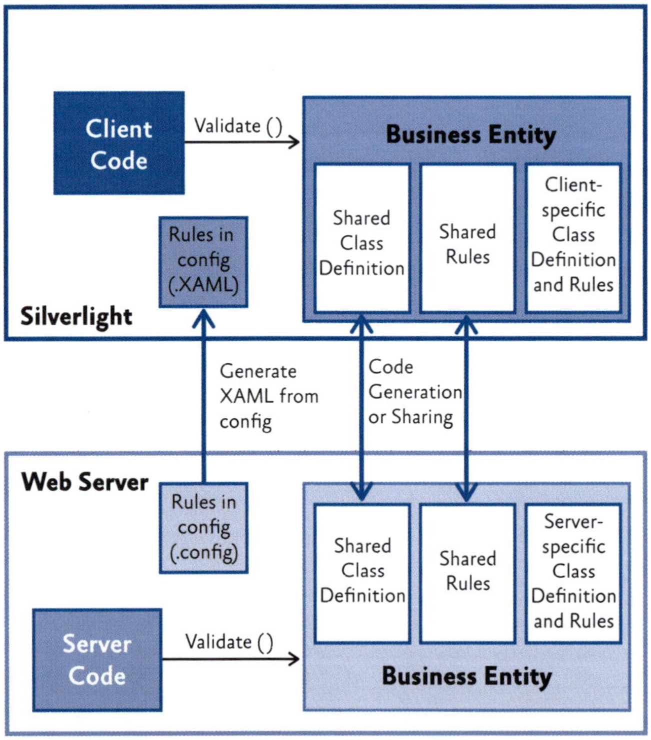

In order to share validation rules between the client and the server, you'll need to have a common definition of the data to validate. Therefore, it's important to have a similar class definition for your business entity on the Silverlight client and on the server. Once you have that common basis, for example certain properties that are the same between Silverlight client's and server's business entities, it becomes possible to share validation rules as well.

When you are sharing validation rules between the Silverlight client and the server, you will need to reference the Enterprise Library 5.0 binaries on the server, but reference the Enterprise Library 5.0 Silverlight Integration Pack binaries on the Silverlight client.

There are several methods for sharing validation rules between the client and the server:

- Linked files
- WCF RIA Services code generation
- Configuration

These options will be discussed further in the following sections.

Sharing Validation Rules Using Linked Files

One method for sharing code between a Silverlight and a non-Silverlight application is to use Microsoft® Visual Studio® development system linked files. This method can be used to share your validation rules, but also to share your class definition between client and server.

We're applying this technique in the AccountHolder Module. We're using WCF Data Services to transfer data between client and server. On the server, we have created special metadata classes for our business entities which contain the validation rules. On the client, WCF Data Services has generated a proxy with the same business entities for us. We're then using Visual Studio linked files to also include the metadata classes in the Silverlight client.

Normally, in Visual Studio all of your projects files are stored under the same folder as the project file itself. Using linked files, you can also use files that are stored outside of the project tree. This way, you can include a file that is physically in a different location.

TO ADD A LINKED FILE:

1. In the solution explorer, open the project where you would like to add the linked file
2. Right click and select **Add existing file**

3. Browse to the file to add and using the drop-down box next to the add button, select **Add as link** as shown in the screenshot below:

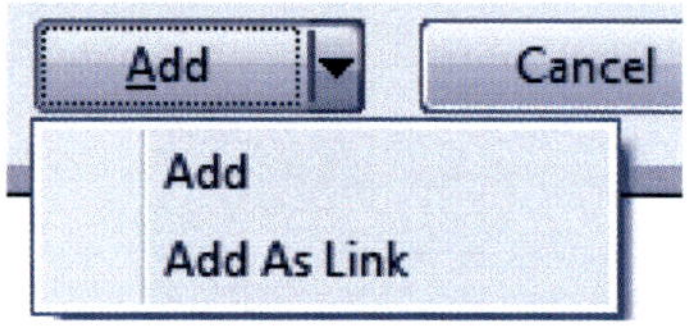

Of course, if you have decorated your business entities with validation attributes, you'll only have to share your class files with your business entities. If you have your validation logic stored in different classes, it's also possible to share those files.

For more information on **adding a linked file**, please refer to MSDN.

Using Business Entities as Data Transfer Objects in WCF

If you are using linked files to share your business entities between the client and server, then it's often also possible to use your business entities as data transfer objects. This is a common approach for building Silverlight applications, since both the client and the server are based on Microsoft technologies.

> **Note:** *This technique should not be used if your web services need to be interoperable with non-Microsoft technologies. A Contract-first approach is then recommended to avoid interoperability issues.*
>
> *This technique is also not recommended if you have many different consumers of the same web service, because you're effectively exposing the internal structure of your application to the outside world. Any changes to your business entities will directly impact your client. This is not so much a problem if the client and the server effectively form a single application, but can become very difficult if you have many different consumers of the same web service. In that case, it's better to consider creating separate data transfer objects that can be versioned independently of the internal workings of your service.*

If you use business entities as data transfer objects in WCF services or WCF Data Services, then you must take special actions when creating a client proxy. The following diagram shows how this would work:

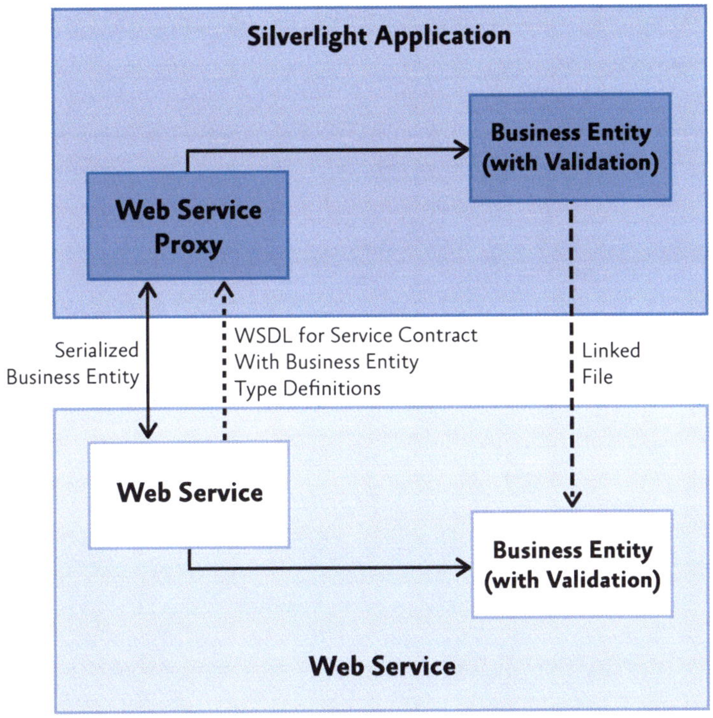

Your web service application defines the business entities that may be decorated with validation attributes. By using the business entities in your service contracts, WCF will generate a WSDL with the type definitions for your business entities. In your Silverlight application, you'll need a web service proxy to access the application, which will be generated by Visual Studio. For regular WCF services, you can reuse the types in referenced assemblies. This means that the generated proxy uses your linked business entities, annotated with validation attributes, rather than generating business entities itself.

> Silverlight and the .NET Framework also allow for limited binary compatibility between Silverlight and .NET Framework assemblies. Initially we also looked at using this to share our business entities between client and server. Unfortunately, this would not work in our situation, because the enterprise library validation assemblies are not binary compatible. The business entities must reference these assemblies, because they use validation attributes to define validation rules, so the assemblies that that hold the business entities cannot be shared either.

Using Partial Classes Or Conditional Compilation

The linked code will be compiled for both the server and the Silverlight client. This means you can only use language constructs and other classes that are available on both platforms. By using partial classes or conditional compilation statements, you can add platform-specific code to your classes, but keep the shared file platform agnostic.

Sharing Validation Rules Using WCF RIA Services

One of the features of WCF RIA Services is the ability to share your business entities and their validation rules between the client and the server.

WCF RIA Services typically revolves around an entity model. In your entity model, you design your business entities and how they map to database tables. Using a code generator, you'll be able to generate business entities on the server. Because the properties of these business entities are generated as well, you cannot directly add validation attributes to them. To solve this, WCF RIA Services allows you to add validation attributes to separate metadata classes. Both the WCF RIA Services and the Validation Application Block validation engines will also read validation attributes from these metadata classes. You can use both WCF RIA Services validation attributes and the Validation Application Block attributes to annotate the business entities or metadata classes.

To share the business entities and their validation logic with the client, WCF RIA Services relies on a code generator. Any business entities on your server that are exposed through your business service will get a generated counterpart on the client. However, these classes don't use metadata classes to define their validation attributes, but get their validation attributes directly generated on their properties.

Please refer to the chapter on **Using the Validation Application Block in a WCF RIA Services application** for more information.

Sharing Validation Rules Using Configuration

Another way you can share validation rules between the client and the server is by using configuration. If you wish to do so, both the client and the server need a common definition of your business entities. The class names, their namespaces, and the properties you wish to validate must be the same on both client and server. As described earlier, this can be achieved by using linked files or by using WCF RIA Services to share your business entities. However, if you use WCF Data Services or regular WCF services to generate a proxy for you, then the business entities generated for that proxy will typically be in a different namespace.

On the server, you can define validation rules in the web.config file, for example by using the configuration console. The Validation Application Block will automatically read any validation rules defined in the web.config file.

In Silverlight, you can define validation rules in a XAML configuration file that is either embedded in the XAP or downloaded from the server. This can be used to share validation rules between the client and the server, because you can generate XAML configuration files from the web.config file. For more information about the XAML generation process, please refer to the **Configuration** chapter.

If you wish to use validation rules defined in a configuration file, then you must trigger validation using the Validation Application Block Validator class. WCF RIA Services will not pick up any validation rules defined in configuration.

In Stock Trader V2, the web.config also contains Enterprise Library configuration that should not be converted to XAML, such as the logging and exception handling settings for the web server. To be able to share the validation rules between the client and the server, we decided to store the validation configuration in a separate file configuration source and only convert that file to XAML. The chapter on **Configuration in the Stock Trader** describes this process in more detail.

Using the Validation Application Block in a WCF RIA Services Application

The Validation Application Block works well with WCF RIA Services validation and can be used to extend WCF RIA Services with support for rule sets and validation rules defined in configuration. There are a couple of caveats to keep in mind, however.

- **Triggering Validation.** Both WCF RIA Services and the Validation Application block provide a validator class that is used to trigger validation. Depending on which technology triggers the validation, you'll get slightly different behavior. We'll discuss these differences in more detail in this section.
- **RIA Services code generation.** WCF RIA Services uses a code generator that reads the server-side business entities with their validation rules to generate business entities with the same business rules on the client. There are some limitations to the attributes that the default code generator can use. We'll also discuss these limitations in this section.

When WCF RIA Services Triggers Validation

The class in WCF RIA Services that performs the validation is called the **System.ComponentModel.DataAnnotations.Validator** class. While it's possible to call this validator class manually, WCF RIA Services will sometimes also call the validator automatically. For example: An entity is automatically validated when it's loaded from the database (to ensure it's valid) and also before you submit any changes to the server.

When the DataAnnotation's **Validator** class is called (either explicitly in your code, or implicitly by WCR RIA Services), it will validate all the DataAnnotations validation attributes, including the Validation Application Block attributes, as the latter are also derived from **System.ComponentModel.DataAnnotations.ValidationAttribute.**

There are some limitations with regards to using the Validation Application Block attributes, however. Since the DataAnnotations Validation class does not allow you to specify a rule set when validating, any Validation Application Block attribute with a rule set specified will be ignored by the **DataAnnotations Validator** class. Also, the **DataAnnotations Validator** class will not look for validation rules defined in configuration, so those validation rules will also be ignored.

If you wish to call validation rules with a specific rule set or call validation rules that are specified in configuration, then you should trigger validation explicitly using the Validation Application Block's **Validation** class.

For more information on WCF RIA Services validation, please refer to the WCF RIA Services online documentation on MSDN.

When the Validation Application Block Triggers Validation

The Validation Application Block also provides a class that can trigger validation, called **Microsoft.Practices.EnterpriseLibrary.Validation.Validator**.

Unlike the **Validator** class from DataAnnotations, this class is not called automatically by WCF RIA Services. You must explicitly call one of the **Validate** methods on the validator to initiate validation.

When you call one of the validate methods on this **Validator** class, it will call all validation attributes that are defined on your class, including those defined by DataAnnotations attributes. It will also read validation rules defined in configuration and execute those. The **Validator** class also allows you to specify a rule set to validate. If you do so, then only validation rules that have that specific rule set defined are called.

The Validation Application Block Validator will also respect the **MetadataTypeAttribute** in Silverlight. This means that you can use the **MetadataTypeAttribute** to add Silverlight specific validation rules that will only be used by the Validation Application Block Validator and will be ignored by the WCF RIA Services Validator.

For more information, please refer to **Validating an Object** in the Enterprise Library documentation.

Limitations of the WCF RIA Services Default Code Generator

The default code generator that ships with WCF RIA Services has a couple of limitations:

- **Incorrect constructor parameters cause attributes to be skipped.** The attributes for the Validation Application Block perform validation of their constructor parameters. For example, if you create an **EnumConversionValidatorAttribute** and you pass null as the enum type, it will throw an argument null exception. The WCF RIA Services code generator will skip any attribute that throws an exception while creating it. This causes a different behavior in Silverlight. On the **web server**, you will get an exception when you create a business entity with an incorrect validation attribute on it. In the Silverlight client, you will not get an exception because the failing attribute has not been generated.
- **DomainValidatorAttribute.** The domain validator attribute allows you to check the value of a property against an array of predefined values. The code generator does not correctly generate attributes with arrays as parameters. To help with that problem, 10 additional constructor parameters have been added. This allows you to specify up to 10 values to validate against.
- **RangeValidatorAttribute does not work properly with decimals and long values.** When the WCF RIA Services code generator generates a Range Validator on the client, it will convert all numeric upperbound and lowerbound values to integers. That means that you can only use the range validator on integers.
- **RegexValidatorAttribute and ValidatorCompositionAttribute.** By default, the WCF RIA Services code generator will not generate full type names for all parameters. This causes issues for the RegexValidator, because it uses the RegexOptions enum from the **System.Text.RegularExpressions** namespace and will cause a compile error in the generated code. This also causes issues for the ValidatorCompositionAttribute, because it uses

the CompositionType enum, which resides in a different namespace than the attribute. To solve this problem, you must instruct the code generator to include the full type names in the code generation. You can do this by adding the following XML snippet to your Silverlight project file:

```
XML
<PropertyGroup>
    <RiaClientUseFullTypeNames>true</RiaClientUseFullTypeNames>
</PropertyGroup>
```

Object Validator Attribute and Object Collection Validator Attribute

You cannot use the **ObjectValidatorAttribute** or the **ObjectCollection Validatorattribute** on a WCF RIA Services business entity directly, without specifying a rule set. These attributes will throw an exception with the message "A validation attribute of type **ObjectValidator Attribute** cannot be used to validate values."

The reason that this exception is thrown is that the Object Validator Attribute and Object Collection Validator Attribute only make sense on the client and wouldn't work as expected on the server.

In the Orders Module for the stock trader, we have two WCF RIA Services business entities: Order and Order Details. On the client, you might want to use the object collection validator so that when you validate the order, its order details get automatically validated as well. The problem is that you might expect the same behavior to occur on the server and that doesn't happen. When you submit an order with order details to the server, you can validate the Order Details objects and the Order object individually, but not as the graph of objects.

If you accept that the Object Collection Validator Attribute and Object Collection Validator only work on the Silverlight client, then you can add them on the client using the **MetadataTypeAttribute**. The following code snippet, from the Stock Trader Order Module demonstrates how to do this:

```
C#
[MetadataType(typeof(Order.OrderMetadata))]
public partial class Order
{
    internal class OrderMetadata
    {
```

```
        // When validating the Order, also validate the
        // OrderDetails
        [ObjectCollectionValidator]
        // when validating with the ruleset "WarningRuleset",
        // validate the orderdetails also with ruleset
        // "WarningRuleset"

        [ObjectCollectionValidator(typeof(OrderDetails),
        "WarningRuleset", Ruleset = "WarningRuleset")]
        public EntityCollection<OrderDetails> OrderDetails
            { get; set; }
    }
}
```

In the Silverlight client, the **Order** class has been generated by WCF RIA Services, with all the validation attributes that where specified in metadata on the server. By using a partial class, you can specify a Silverlight-specific **OrderMetadata** class, which contains additional validation rules that will only be used by the Validation Application Block in Silverlight.

Performing Server-Side Validation with Rule Sets or Validation Rules Defined in Configuration

When you submit a WCF RIA Services business entity to the server, it will automatically be validated by WCF RIA Services using the DataAnnotations Validator class. However, DataAnnotations doesn't support rule sets or validation rules that are defined in configuration, so these validation rules will not be executed automatically.

If you wish to execute validation rules that use rule sets or are defined in configuration, you will need to explicitly use the Validation Application Block Validator. If a validation error occurs, then you will need to throw a **ValidationException**, from the **System.ComponentModel.DataAnnotations** namespace. In Silverlight, this will be picked up by WCF RIA Services and it will set the Validation Result on the correct object instance.

The Stock Trader contains a helper class to help with this server-side validation:

C#

```
public class EnterpriseLibraryServerValidator
{
    public void Validate(object objectToValidate)
    {
        Validate(objectToValidate, string.Empty);
```

```
    }
    public void Validate(object objectToValidate, string ruleset)
    {
        IEnumerable<ValidationResult> validationResults =
            GetValidationResults(objectToValidate, ruleset);
        if (validationResults.Any())
        {
            throw new ValidationException(validationResults.
                First(), null, objectToValidate);
        }
    }
    public IEnumerable<ValidationResult> GetValidationResults(
        object objectToValidate)
    {
        return GetValidationResults(objectToValidate, string.
            Empty);
    }
    public IEnumerable<ValidationResult> GetValidationResults(
        object objectToValidate, string ruleset)
    {
        var validator = ValidationFactory.CreateValidator(
            objectToValidate.GetType(), ruleset);
        var entlibValidationResults = validator.
            Validate(objectToValidate);
        return TranslateToDataNotationsValidationResult(
            entlibValidationResults);
    }
    IEnumerable<ValidationResult>
        TranslateToDataNotationsValidationResult(
        ValidationResults entlibValidationResults)
    {
        foreach(var result in entlibValidationResults)
        {
            yield return new ValidationResult(result.Message, new
                []{result.Key});
        }
    }

}
```

This class, called the **EnterpriseLibraryServerValidator** will help you to trigger the Validation Application Block validation, translate any validation results to a list of Data Annotations Validation Results and then throw the correct type of exception. It will also allow you to pass a rule set to execute conditional validation rules.

Using the Validation Application Block in a WCF Data Services Application

WCF Data Services is a component of the .NET Framework that enables you to create services that use the Open Data Protocol (OData) to expose and consume data over the Web or intranet by using the semantics of representational state transfer (REST). OData exposes data as resources that are addressable by URIs. Data is accessed and changed by using standard HTTP verbs of GET, PUT, POST, and DELETE. OData uses the entity-relationship conventions of the Entity Data Model to expose resources as sets of entities that are related by associations.

In Silverlight, you can also consume OData services. The Silverlight WCF Data Services components can generate a proxy for you that will contain a client-side representation of the business entities exposed by the OData Service.

WCF Data Services doesn't have built in support for validation, like WCF RIA Services, but you can use Enterprise Library to implement validation both on the Silverlight client and / or the server.

Using the Metadata Type to Define Validation Rules

Both on the server and on the Silverlight client, by default, WCF Data Services will generate your business entities for you. That means that you cannot add validation attributes to your business entities directly, but must add them to metadata classes that you have defined using the MetadataType attribute.

The following code sample, from the Account Holder module in Stock Trader V2, demonstrates how you can add validation rules to the AccountHolder class in meta data:

```
C#
[MetadataType(typeof(AccountHolderMetadata))]
public partial class AccountHolder
{
   [HasSelfValidation]
   private class AccountHolderMetadata
   {
      [StringLengthValidator(1, RangeBoundaryType.Inclusive, 0,
         RangeBoundaryType.Ignore, MessageTemplate = "The first
         name is required")]
      public string FirstName { get; set; }
   }
}
```

In the Stock Trader V2, we're using linked files to share these validation rules between the Silverlight client and the server. We did need to make a couple of tweaks to get that to work. We had to make sure the namespace used for the AccountHolder class is the same in both the client and server. Also, the MetadataType attribute in Silverlight is located in a different namespace, as it is provided by the Validation Application Block and not by the built-in **DataAnnotations** namespace. We added conditional compilation instructions to solve this problem, so the correct namespaces are used.

Performing Server-side Validation

When you wish to perform validation on the server and propagate these validation results to the client so you can display them in the user interface, then there are three steps you'll need to perform:

- **On the server, add Change Interceptors to trigger validation.** The way that WCF Data Services allow you to implement custom validation is through methods that are decorated with the **ChangeInterceptor** attribute. To get a validation error to propagate from the server to the client, you will need to encapsulate it in a DataServiceException instance. Stock Trader V2 shows how you can do this using a helper called the **Data ServicesValidationHelper**. The following example, from the **AccountHolderService** shows how this is implemented in the Stock Trader:

```
C#
[ChangeInterceptor("BankAccountSet")]
public void OnChangeBankAccount(BankAccount account,
   UpdateOperations operations)
{
   /// ...
   DataServicesValidationHelper.Validate(account, operations);
   if (account.Number == "123456")
   {
      var results = new ValidationResults();
      results.AddResult(new ValidationResult("Our records
         indicate that you are not the owner of bankaccount
         with number '123456' (Server side validation rule)",
         account, "Number", null, null));
      DataServicesValidationHelper.ThrowDataServiceException(
         results);
   }
}
```

- **On the client, parse the exception to retrieve the Validation Results.** When an exception occurs on the server, WCF Data Services will serialize the exception to XML. When the exception reaches the client, you will need to parse the XML to retrieve the exception that occurred on the server. The Stock Trader V2 application also contains a **DataServicesValidationHelper** Silverlight class to help with this parsing. The following example shows how you can get use the **DataServicesValidationHelper** to extract validation results from a data services exception.

C#
```
ValidationResults validationResults;
if (DataServicesValidationHelper.TryGetValidationResults(
   dataServicesException, out validationResults))
{
   return new ValidationErrorException(validationResults);
}
else
{
   return dataServicesException;
}
```

Implementing Validation in the Stock Trader

There are two places where validation has been implemented in the Stock Trader V2. The Account Holder module was built using WCF Data Services and the Orders module was built using WCF RIA Services. In this topic, we'll describe how these modules have been built.

While building the Account Holder and Order Module, we found that we had to write a lot of repetitive code to build our view models. There are a couple of interfaces that you always have to implement, such as **INotifyPropertyChanged** and **INotifyDataErrorInfo**. To minimize the repetitive code, we have built some infrastructure pieces. They help with the most common scenarios for building views / view models. These infrastructure pieces will be discussed in more detail in the section called Advanced (View) Model Scenarios.

Validation with WCF Data Services (The Account Holder Info Module)

The Stock Trader V2 allows its users to change their own account information through the Account Holder Info Module. In this module, users can register certain information, such as their name or birth date. It also allows users to enter one or more bank accounts.

Account Holder Info Module Scenario

Most of the functionality for the Account Holder Info Module was already present in the previous version of Stock Trader, so one of the most important goals for the new Account Holder Info Module was to reuse the database schema, business entities and validation logic.

Two years ago, we invested heavily in improving the account holder functionality in the previous Stock Trader so that it used Entity framework and the Validation Application Block for defining validation rules on its business entities. We couldn't afford to rebuild this logic from scratch, so we were very happy that we were able to reuse that logic when building Stock Trader V2.

Design of the Account Holder Module

The following diagram shows the data model for the Account Holder Info Module. As you can see, an account holder can have several bank accounts. There must be a single primary bank account.

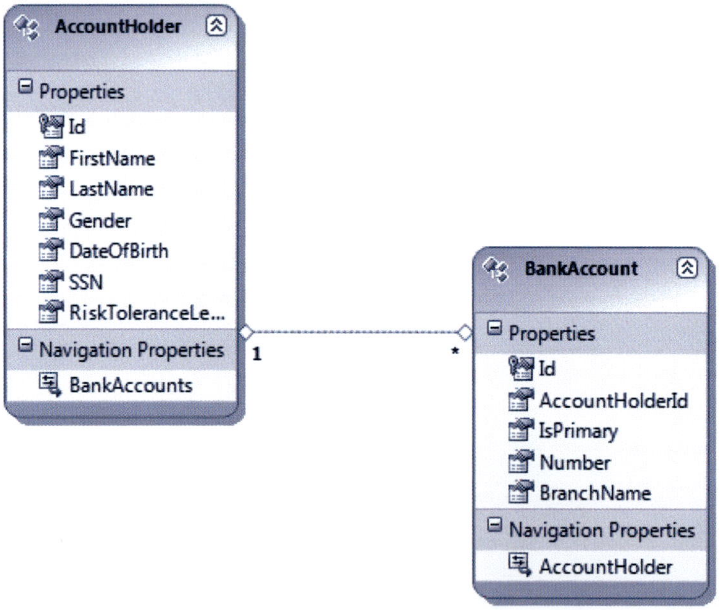

In most examples online, you'll only see a single object being validated. In our case, that wasn't enough. We wanted to be able to validate an account holder, including the associated bank accounts information, as a whole. There are some challenges that arise from having to validate a graph of objects, such as where to store the validation results. This is described in more detail in the section called **Advanced ViewModel Scenarios**.

The following diagram shows a schematic representation of the design of the Account Holder Info Module.

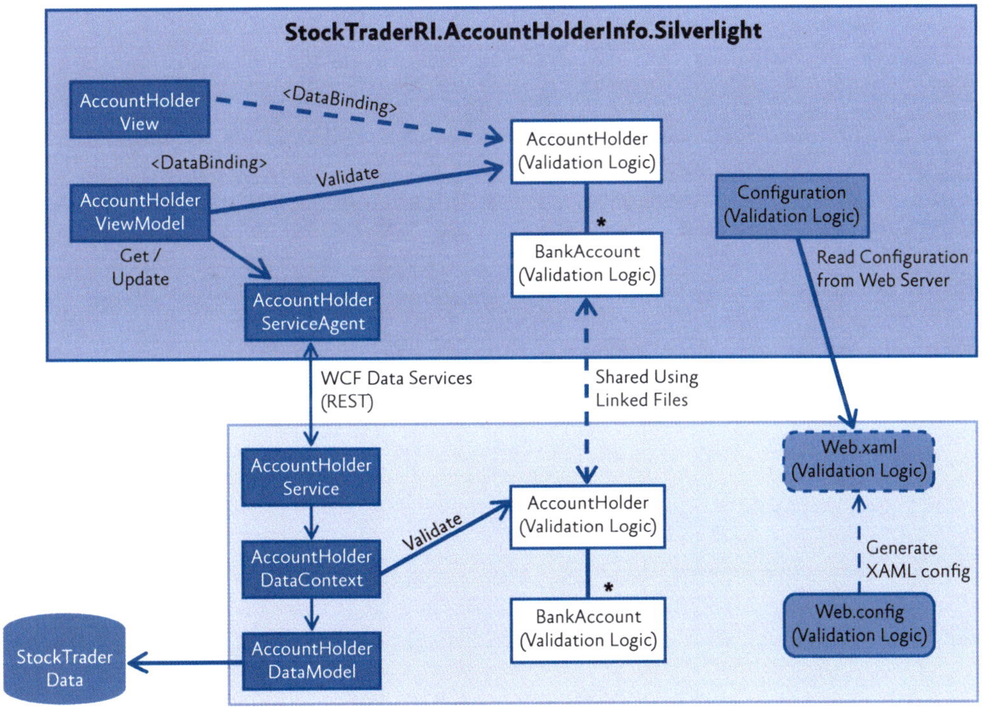

The Account Holder Info Module is built as a set of WCF Data Services that expose two business entities: **AccountHolder** and **BankAccount**. An Entity Framework data model called **Account HolderDataModel** is used to persist the business entities in the **StockTraderData** database.

We could have built the Account Holder Info Module as more traditional WCF services, but we decided not to do that because in the future, we expect other clients to also use these web services. Since these other clients prefer to communicate using a REST API, WCF Data Services seemed a really good candidate.

The Silverlight client is built using the Model View ViewModel (MVVM) pattern. It communicates with the web services through a **ServiceAgent** class.

We have designed the Account Holder Info Module to be easily unit tested. For example, by isolating the UI Logic in the Account Holder View Model and isolating all data access logic in the AccountHolder ServiceAgent, we could write automated unit tests to virtually all UI logic without having to do UI automation tests.

The logic for order processing in the old Stock Trader was not pretty, and its validation logic was even worse. Instead of using a well-designed validation framework, it was a big pile of inconsistent spaghetti code, with lots of duplication. When we started the work on Stock Trader V2, we quickly decided that the best approach was to throw away the old code and start over.

Most of the validation rules are defined as attributes on the business entities. Since the validation rules must be run on the server (for security purposes) but also on the client (to improve the user experience), we have used Visual Studio linked files to share the business entities between the client and the server.

Validation with WCF RIA Services (the Orders Module)

In the Stock Trader V2 Reference Implementation, the **Orders** module has been built using both WCF RIA Services and the Validation Application Block.

Orders Module Scenario

The previous version of Stock Trader also had order processing functionality. However, it was very complicated and hard to maintain.

The Orders module for the new Stock Trader V2 has been reworked from scratch. WCF RIA Services was chosen as the platform, largely because of the built-in validation support, and rich Microsoft® Visual Studio® tooling available in WCF RIA Services.

While most of the validation logic could easily be built using the WCF RIA Services built-in validation support, we did need some of the extra validation support that only the Enterprise Library Validation Application Block provides.

For example, the business wanted the ability to experiment with the cost model associated with placing small orders. If an order was placed for penny stock or for less than a certain number of stocks, they wanted to be able to charge an additional fee. But we cannot charge additional fees without warning the users first. Therefore, we decided to implement this using configurable validation rules in in a specific rule set, called "Warnings."

Design of the Orders Module

The **Orders** Module has the following Data Model: An Order is a logical grouping of several stock transactions that can be managed together or individually.

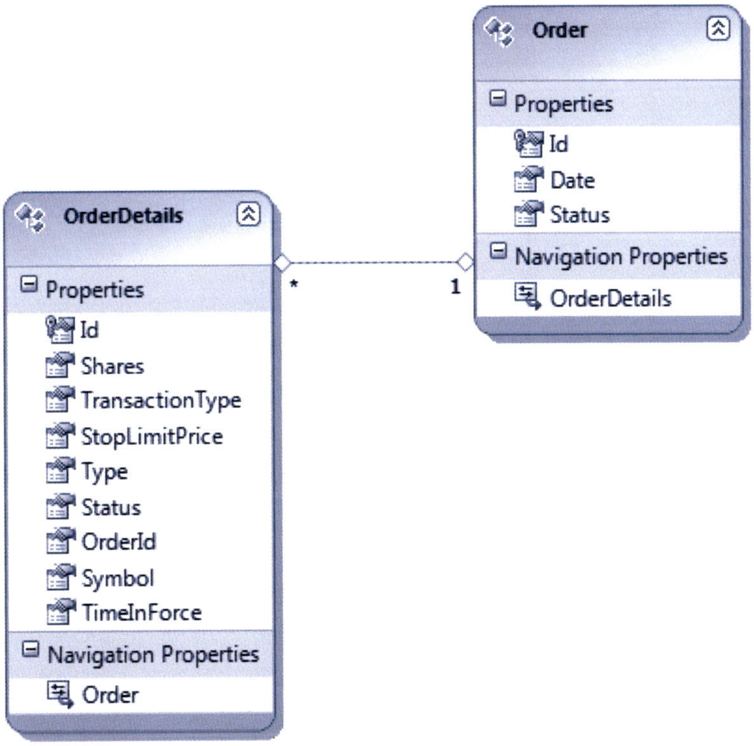

The following diagram explains the **Orders** module in more depth.

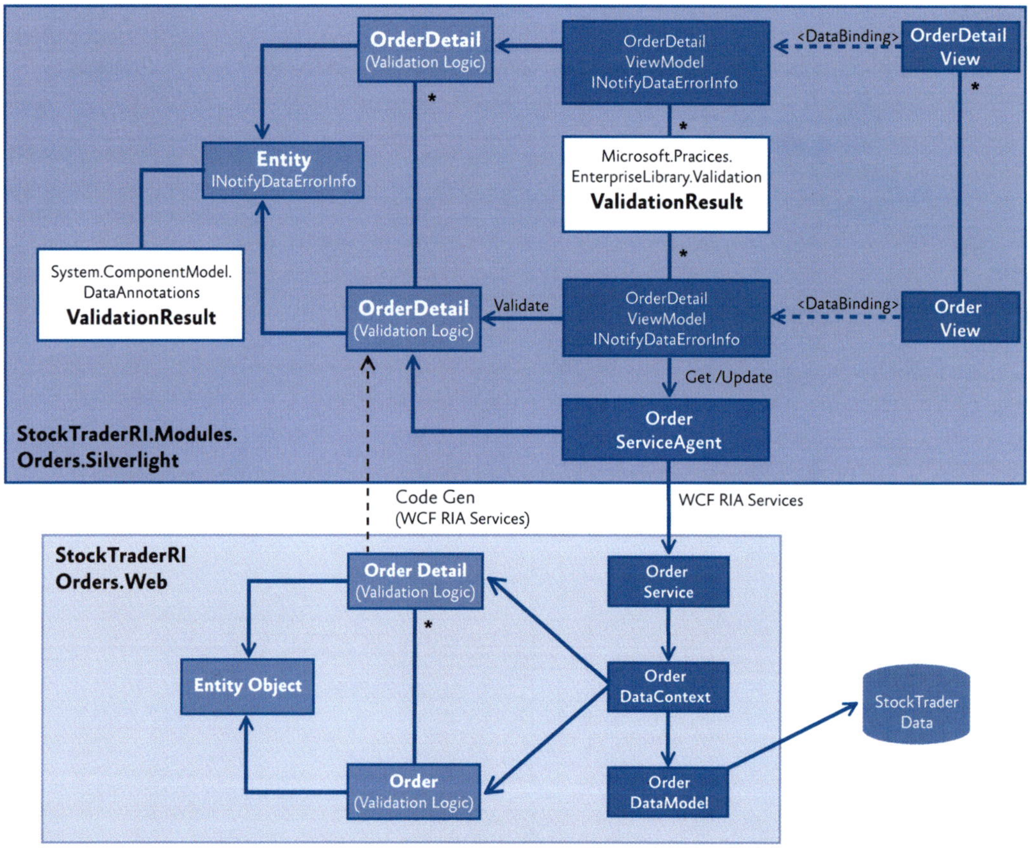

The heart of the **Orders** module is the Order Data Model, built using WCF RIA Services. The business entities, **Order** and **Order Detail**, are generated from this data model. Both have validation rules defined in metadata classes, using a mixture of Validation Application Block validation attributes and DataAnnotations built-in attributes.

WCF RIA Services has also generated the business entities on the Silverlight client. These business entities are also decorated with the validation attributes that were defined on the attributes on the Server.

To implement the UI for order entry, the MVVM pattern is used. The **Order** and **OrderDetail** business entities are only responsible for containing the data, which are retrieved and stored through the **OrderServiceAgent**. The UI logic is isolated in the **OrderViewModel** and **OrderDetailViewModel** classes, so these classes can be unit tested in isolation. The presentation is handled in the **OrderView** which uses zero or more **OrderDetailViews**.

Both the **Order** and **OrderDetail** classes in Silverlight inherit from the WCF RIA Services Entity class. Therefore, they already implement **INotifyPropertyChanged** and **INotifyDataErrorInfo** and expose a collection of (DataAnnotations) validation results. This makes it very easy to bind these classes to the UI and show validation errors.

Sequence for Validation

There are several points in time when validation is triggered:

1. When the user changes a value in a textbox, the bound property is automatically validated. For the WCF RIA Services entities, this is done automatically by the generated code. For the view models, this is done calling the **ValidateOnPropertyChanged** method on the Model Graph Validator.
2. When you press the submit button, the Validation Application Block's validation is triggered, with the default rule set. This will also trigger the validation of the entire Order and OrderDetails object graph, because the OrderViewModel has an **Order** property with the **ObjectValidator** attribute on it. The Order has a collection of Order Details with the **ObjectCollectionValidator** attribute (defined in the **OrderMetadata** class) on it.
3. When the order is valid, then a second validation pass is executed. This time with the 'Warning' rule set. Only the Validation Application Block's validation attributes or attributes defined in configuration can be used here, because DataAnnotations' built-in attributes do not support rule sets. If a validation warning occurs, then we'll show the validation warnings and give the user the chance to stop.
4. When the order is submitted, WCF RIA Services will automatically trigger validation using the DataAnnotations validator on the server. This will validate using all the DataAnnotations built-in validation attributes and also the Validation Application Block attributes that do not have a rule set specified on them.
5. On the server, we'll also explicitly call Validation Application Block validation again. We're also executing a 'server only' validation rule, which guards against placing an order that's larger than your risk profile. If you place an order that's greater than 1025 stocks, this will raise a validation error.

Advanced (View) Model Scenarios

When you are building a user interface using the MVVM pattern, you'll quickly find that there are several repetitive tasks to be performed. For example, to create a model or view model that is easily bindable to the UI, you should implement the **INotifyProperty Changed** interface and raise this event in every property setter. Then to support validation, you'll want to implement the **INotifyData ErrorInfo** interface.

To make implementing these interfaces easier, a custom library called StockTraderRI.Infrastructure.ModelSupport was developed. This library supports the most common scenarios when building models and view models that should be easily data bound to the user interface.

This chapter stars with an overview of the Model Support library and will then describe the individual scenarios that it supports.

Design of the StockTraderRI.Infrastructure.ModelSupport Library

The ModelSupport Library was designed like the layers of an onion. If you want to get started quickly, you can take the whole onion, but you can also peel back the layers and only use the pieces that you like.

The following diagram shows the most important pieces of the library:

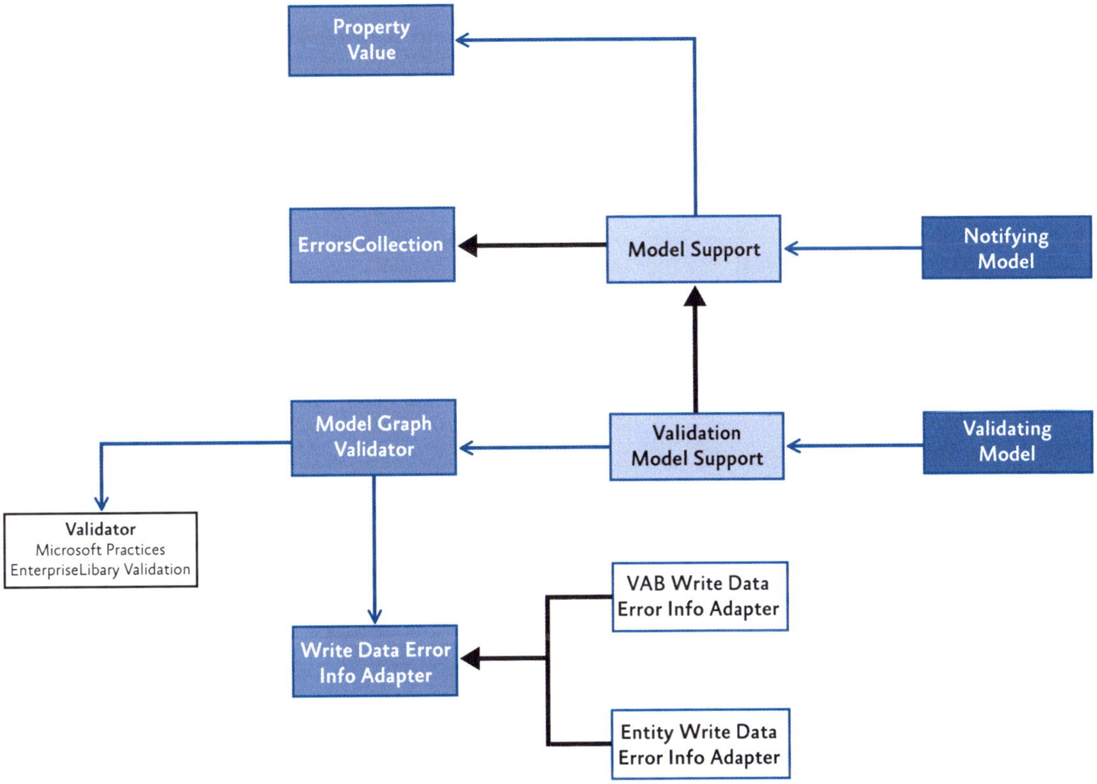

On the left of the diagram are the three core pieces:

- The **PropertyValue** class helps you to implement the **INotify PropertyChanged** interface. It can help you implement quasi-automatic properties, which raise the property changed event when they are changed. But if you want more control, you can also use it just to raise the property changed event explicitly.
- The **ErrorsCollection** class helps you to implement the **INotify DataErrorInfo** interface. It is a collection of validation errors with properties and methods to match the **INotifyDataError Info** interface.
- The **ModelGraphValidator** class helps you to validate a graph of objects. It uses a pluggable mechanism for writing validation errors into models, so that these validation errors easily bind to the user interface. The model validator uses the **Validator** class from the Validation Application Block to execute the actual validation logic.

Using the base classes provided in the Model Support library is a very fast way to get started with building models and view models. However, if you don't want to use a base class or need to use a different base class, then you can still use either the support classes or the core pieces to help you implement the most common scenarios.

In the middle there are two support classes:

- **ModelSupport** combines the functionality for the Property Support and Errors Collection. Since it is very common to have to implement both the **INotifyPropertyChanged** and **INotify DataErrorInfo** interfaces on your models, it would again be very repetitive to have to use both these classes on each model or view model.
- **ValidatingModelSupport** extends the functionality of the **ModelSupport** class with the ability to validate a graph of objects. This helps if you wish to implement validation in your ViewModels, but do not wish to have to explicitly add the **ModelGraphValidator** to each ViewModel.
- On the right, there are two base classes. These base classes have the same functionality as their Support counterparts. In fact they use them internally, but they expose their functionality through methods on the base class.
- The **NotifyingModel** is an abstract base class that uses the **ModelSupport** class to implement **INotifyDataErrorInfo** and **INotifyPropertyChanged**.
- The **ValidatingModel** is an abstract base class that extends the functionality for the Notifying Model with validation support. It uses the Validating Model Support class to implement that support. It also implements the custom **IWriteDataErrorInfo** interface for writing error information into the model.

Implementing the INotifyPropertyChanged Interface Using the Property Support Class

The **INotifyPropertyChanged** interface is used by the Silverlight data binding engine to detect if a property that is bound to the view has changed, so that the view can be updated. The **PropertyValue** class can help you to implement **INotifyPropertyChanged** in your models. The following example demonstrates how to use the **PropertyValue** class:

C#

```
public class ExampleModel : INotifyPropertyChanged
{
    private PropertyValue propertyValue;
```

```
    public ExampleModel()
    {
        this.propertyValue = new PropertyValue(()
        => PropertyChanged);
    }

    private int exampleProperty;
    public int ExampleProperty
    {
        get { return exampleProperty; }
        set
        {
            if (exampleProperty == value)
                return;

            exampleProperty = value;

            propertyValue.RaisePropertyChanged(
                () => ExamplePropertyUsingExpression);
        }
    }

    public int ExampleAutomaticProperty
    {
        get { return propertyValue.GetProperty(()
        => ExampleAutomaticProperty); }
        set { propertyValue.SetProperty(()
        => ExampleAutomaticProperty, value); }
    }

    public event PropertyChangedEventHandler PropertyChanged;
}
```

In the constructor, an instance of the **PropertyValue** class is created, and it gets a lambda expression that retrieves the **Property Changed** event when it needs to be raised. Normally, you can't pass an event to a different class, but you can pass a lambda expression that retrieves the event handler when it's needed.

The first example property shows a regular property that uses the **PropertyValue** class to raise the **PropertyChanged** event. You can either pass a string with the name of the property, or pass an expression that targets the property.

In the beginning I had to get used to writing all these lambda expressions, as opposed to specifying strings for the property names when implementing the **INotifyPropertyChanged** interface without this infrastructure. However, once I got used to them, I found that they offer a lot of advantages. You get compile-time checking for the property name and type and you'll get much better refactoring support within Visual Studio.

Having to implement properties as shown with the ExampleProperty sample above still requires a lot of coding. You have to have a backing variable for the property, check if the property has changed, update the value, and raise an event. You will need eight lines of code for each property. To reduce that amount of code, you can also use quasi-automatic properties, as shown with the ExampleAutomaticProperty example above. The **PropertyValue** class acts as a backing store for the property values, so you don't have to create a private variable. The **SetProperty** method will check if the property has changed and will only raise the **NotifyPropertyChanged** event if it has changed.

The data binding engine will only pick up validation errors from objects that implement **INotifyData ErrorInfo**. If you use a graph of objects, such as the **AccountHolder** and **BankAccount** classes from the Account Holder Info Module, then both these classes need to implement **INotifyDataErrorInfo**.

Implementing INotifyDataErrorInfo Using the Error Collection Class

The **INotifyDataErrorInfo** interface is used by Silverlight to notify the data binding engine when validation errors have occurred. The data binding engine will then query the model for the validation errors and show the validation errors in the view.

You can use the **ErrorsCollection** class to easily implement **INofity DataErrorInfo** on your models. The following code example shows how this works:

C#
```
public class ExampleINotifyDataErrorInfoModel :
   INotifyDataError-Info
{
   private ErrorsCollection<string> validationErrors;
   public ExampleINotifyDataErrorInfoModel()
   {
      validationErrors = new ErrorsCollection<string>(
         this, () => ErrorsChanged);
   }
   public void AddValidationErrorExample()
   {
      this.validationErrors.SetErrors(() => BoundProperty,
         new []{"Error has occurred"});
   }
   public string BoundProperty { get; set; }
   public IEnumerable GetErrors(string propertyName)
   {
      return validationErrors.GetErrors(propertyName);
   }
```

```
    public bool HasErrors
    {
        get { return validationErrors.HasErrors; }
    }
    public event EventHandler<DataErrorsChangedEventArgs>
        ErrorsChanged;
}
```

The **INotifyDataErrorInfo** interface makes no assumption on what a validation error is, so you can use anything as a validation error. In this example, we're using a string to represent a validation error.

The **ErrorsCollection** class is created in the constructor. Using a lambda expression, the **ErrorsChanged** event is passed to the **ErrorsCollection** so that it can raise the **ErrorsChanged** event when validation errors are added or removed.

Using the **SetErrors** method, you can add or remove validation errors for a particular property.

Validating a Graph of Objects Using the ModelGraphValidator Class

The **ModelGraphValidator** class helps you to validate a graph of objects and to assign the validation errors to the object that has the actual error. For example, if you have a Person class that has a property called Address classes and a collection of PhoneNumbers classes, then you might want to validate the entire graph of objects, rather than each object individually.

When you call the **Validator** class from the Validation Application Block on this graph, it will return a list of validation errors, without changing the objects being validated. If you are only validating a single object, you can simply populate the errors collection for that object with the validation result. However, if you are validating a graph of objects, then this approach does not work. The **INotifyDataError Info** interface has to be implemented on all objects that are bound to the user interface and have their error information displayed.

The **INotifyDataErrorInfo** is a read-only interface. It only supports methods for reading error information from that object, but does not define how you can populate your models with validation errors. To support writing validation errors into objects, the Model Support library also contains the custom **IWriteDataErrorInfo** interface. This interface allows you to externally add validation errors into an object so that they can be displayed properly in a view.

The following example shows how you can validate a graph of objects:

```
C#
public class PersonModel : NotifyingModel
{
   private ModelGraphValidator modelGraphValidator;
      public PersonModel()
   {
      this.modelGraphValidator = new ModelGraphValidator(
         this);
   }
   public void Save()
   {
      var result = this.modelGraphValidator.Validate();
      if (!result.Any())
      {
         // perform the actual save operation
      }
   }
   [ObjectValidator]
   public AddressModel Address
   {
      get { return GetProperty(() => Address); }
      set { SetProperty(() => Address, value); }
   }
   [ObjectCollectionValidator]
   public IEnumerable<PhoneNumberModel> PhoneNumbers
   {
      get { return GetProperty(() => PhoneNumbers); }
      set { SetProperty(() => PhoneNumbers, value); }
   }
}
public class AddressModel : NotifyingModel
{
   [StringLengthValidator(0, 100)]
   public string StreetName
   {
      get { return GetProperty(() => StreetName); }
      set { SetProperty(() => StreetName, value); }
   }
}
public class PhoneNumberModel : NotifyingModel
{
```

```
    [StringLengthValidator(0, 3)]
    public string AreaCode
    {
        get { return GetProperty(() => AreaCode); }
        set { SetProperty(() => AreaCode, value); }
    }
}
```

In this example, a person has one address and several phone numbers. For simplicity, we have used the **NotifyingModel** base class to implement **INotifyDataErrorInfo** and **IWriteDataErrorInfo** on these classes, but we could have also implemented these interfaces without using a base class. In order for a validation error to be displayed in the User Interface, it must be added to the error collection of the correct object. For example, if the second phone number for a user has a validation error, then the validation error needs to be added to the second phone number instance. You can't add it to the Person Model's error collection, because then the validation error will not display in the user interface.

Now when you call the save method, you would like to validate the person, but also his address and his phone numbers. You can use the **Object Validator** attribute and **Object Collection Validator** attribute from the Validation Application Block to ensure that these classes also get validated when you call validate.

When you call validate on the Model Graph Validator, it will first retrieve the list of validation errors for the entire graph. Then it will distribute those validation errors to the correct instances so that they display correctly in the user interface.

Using the Model Graph Validator with WCF RIA Services

You can also use the Model Graph Validator to validate business entities generated by WCF RIA Services. The business entities generated by the WCF RIA Service code generator derive from the Entity class. This class implements **INotifyPropertyChanged** and **INotifyDataErrorInfo** for you. It also exposes a collection of validation errors.

The Model Graph Validator uses adapters to write validation errors into the models. There is an adapter for classes that implement **IWriteDataErrorInfo**, such as the **NotifyingModel** and an adapter for classes that derive from the WCF RIA Services **Entity** class.

This feature is very useful when you're building View Models that use WCF RIA Services entities. To implement the view models, you can use **IWriteDataErrorInfo** and validate them together with the WCF RIA Services entities.

Automatic Validation When a Property Changes

It's usually a good idea to validate a user's input as soon as he leaves the input field. This will give the user early feedback about potential validation errors, which often results in a better user experience.

To avoid having to add validation logic to each property setter, you can use the **ValidateOnPropertyChanged** method from the Model Graph Validator. If your model implements **INotifyProperty Changed**, then you can use this to trigger validation as soon as your properties change.

Displaying Validation Errors That Are Not Bound in the UI

The Silverlight Control Toolkit ships with a very useful control called the **ValidationSummary**. This control will pick up any validation error that is bound to the user interface and display them in a summary. This works great for properties that are actually bound to the user interface. However, it's not uncommon for some validation errors to occur that are not bound to the user interface.

> In the Account Holder Info Module, when you validate an account holder, there are certain validation rules that are not bound to any particular piece of the user interface. For example the rule that you should have at least one primary bank account. You can't say: "this control that's bound to this property of my model is invalid," because it's more a problem of the state of the entire object graph, rather than a single property that has an error.

To ensure that validation errors that are not bound to the UI are also displayed in the **ValidationSummary**, a custom behavior was created, called the **ValidationResultsSyncBehavior**. When the validation errors collection of the data context changes, it will find the validation error messages have not yet been added automatically to the validation summary and add them explicitly.

The following code sample, from the EditAccountHolderView, demonstrates how you can add a validation summary control with the Validation Results Sync Behavior:

XML
```xml
<sdk:ValidationSummary>
   <i:Interaction.Behaviors>
      <Validation:ValidationResultsSyncBehavior
         Source="{Binding}" />
   </i:Interaction.Behaviors>
</sdk:ValidationSummary>
```

Multi-Threading Support

All the classes in the model support library help to make it easier to create multi-threaded applications. Silverlight does not allow you to manipulate the user interface from any thread other than the UI thread. If you have a model that's bound to the user interface, and you change one of its properties from a different thread, this also causes an indirect user interface manipulation from a different thread, resulting in an exception.

For this reason, all the classes in the Model Support library will raise their events on the UI thread. The most important reason these events exist is to update the user interface. For example the **Property Changed** event from **INotifyDataErrorInfo** is often used to notify the view of changed data.

So if you use the **Property Value** class to implement **INotify PropertyChanged**, it will always trigger the **PropertyChanged** event on the UI thread, regardless of from which thread you made the change. While you will still need to watch out for the typical multi-threaded issues, such as race conditions or deadlocks, this can make multi-threaded programming a little easier.

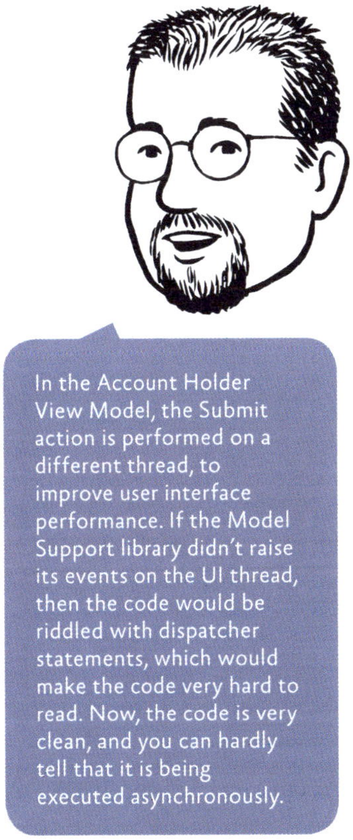

7 Caching

Caching is a programming technique that, when used properly, can help you improve the performance and startup time of your application. Caching allows you to store data that is frequently used or costly to retrieve so that it can be accessed faster.

In this chapter, we'll take a look at the new Caching Application Block from the Enterprise Library Silverlight Integration Pack. Since this is a new application block that does not build upon the caching application block from Enterprise Library 5.0, we'll also look at the API that this block offers. We'll look at how different caching strategies, such as reactive and proactive caching, were implemented in the Stock Trader V2 reference implementation. At the end of this chapter, we'll show various examples of how to use the Caching Application Block.

Caching Using the Caching Application Block

The new Caching Application Block provides a consistent API for adding and retrieving items from a cache. The following code shows how you can use the cache from C#.

C#

```
object item = cache["Item"];
if (item == null)
{
   item = CreateItem();
   cache.Set("Item", item,
      new CacheItemPolicy { AbsoluteExpiration =
      DateTime.Now.AddHours(2) });
}
```

Using the configuration editor, you can define what type of cache you'd like to use. The Caching Application Block allows you to store cached values in memory or in isolated storage:

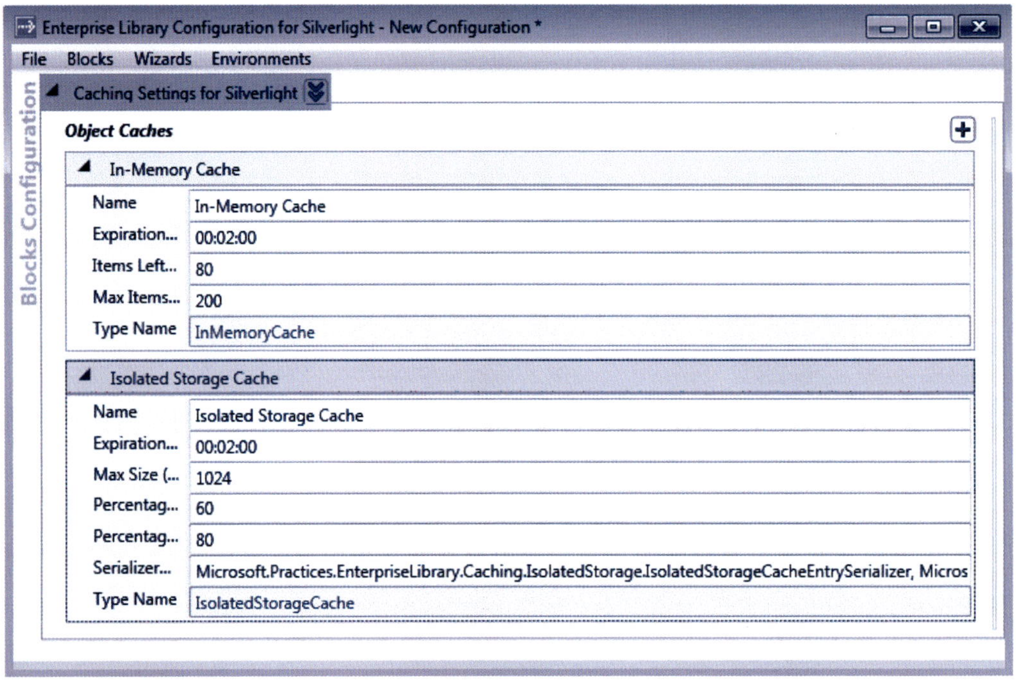

Caching Considerations in a Silverlight Application

If you decide to apply caching in a Microsoft® Silverlight® browser plug-in application, there are several things you'll need to take into consideration.

Memory Is Not Shared with Other Users

Caching in a Silverlight application is somewhat different from caching in a web application. In a web application, the memory of the server and the cached data is typically shared by potentially hundreds of users. Because of that, you'll prefer to cache items that are used by many users. You'll also want to limit the size of the cache, because server memory is also needed to handle requests from all the users.

A Silverlight application is only used by a single user. This means that you don't have to share the client computer's memory with potentially hundreds of users. Unless your application works with really large data sets, the size of the cache in memory is usually not a constraint.

Isolated Storage Size Limit

In a normal Silverlight application, if you wish to store cached data in a persistent way, you can store it in isolated storage. By default, the isolated storage quota is limited to 1 Mb per domain. This means that the isolated storage quota is shared between all applications downloaded from a particular domain name, such as www.microsoft.com. 1 Mb is usually enough space just for caching, but if you decide to use isolated storage to store other information, such as log information, then you can quickly run out of space. While it is possible to increase the size of isolated storage, this must be explicitly allowed by the end user. In a corporate environment, it is also possible to increase the size of the quota for a particular application by using a Microsoft .NET Framework security policy.

For most Silverlight applications, this means that you cannot rely on the fact that the isolated storage quota has been increased. If your application needs more space in isolated storage, then you should make this clear to your users (so they won't decide to deny your application extra space) but you should also ensure that your application will work with minimal isolated storage space.

For more information about isolated storage, please read this topic on MSDN®.

Browser Caching

For certain types of web requests, the result is cached by the browser cache. For example, the **WebClient** class uses the browser cache. On the server you can use the **System.Web.HttpCachePolicy** class to specify how a particular page should be cached by the browser.

Not all server requests are cached by the browser. For example, requests made by Windows® Communication Foundation (WCF), WCF RIA Services or WCF Data Services bypass the browser cache. In those cases, you'll need to cache the results manually.

So for certain types of requests, you can use the browser cache to give you a transparent, server-controlled caching mechanism.

Server-Controlled Cache Expiration Is Difficult to Implement

Silverlight applications usually don't have a continuous connection to the server. This makes it difficult for a Silverlight client to detect when something has changed on the server that should invalidate certain values in the Silverlight application's cache. For example, in ASP.NET it is possible to use the Microsoft SQL Server® cache dependency to invalidate a cached item when the underlying data has changed.

In Silverlight this would be difficult to achieve, because the web server would then need to notify all its clients that a particular piece of data has changed.

Silverlight Caching Application Block

The Caching Application Block in the Enterprise Library 5.0 Silverlight Integration Pack is an entirely new implementation, which very closely resembles the **System.Runtime.Caching** API available in .NET Framework 4.0.

This new implementation provides the following features:

- In-memory cache
- Persistent cache support in isolated storage
- Various cache item purging strategies based on available size in isolated storage, maximum number of cached items, or expiration delay
- These features will be discussed later in this chapter.

Comparing to Enterprise Library Caching Application Block

The caching support in the Enterprise Library 5.0 Silverlight Integration Pack is not based on the Enterprise Library 5.0 Caching Application Block. The main reason for this is the fact that the Caching Application Block is going to be deprecated in the next version of Enterprise Library, in favor of using **System.Runtime.Caching**.

The main scenarios that the Enterprise Library 5.0 version of the Caching Application Block support do not make much sense in a Silverlight application for a number of reasons. For example, storing cached data in a database is not possible in Silverlight.

Future versions of Enterprise Library will also build on top of **System.Runtime.Caching**.

System.Runtime.Caching

The .NET Framework 4.0 introduced the **System.Runtime.Caching** namespace. This namespace contained a new and extensible caching implementation. For example, this implementation allows you to create custom caching providers.

In Silverlight 4, the **System.Runtime.Caching** namespace does not yet exist. The Enterprise Library 5.0 Silverlight Integration Pack contains classes that are very similar to **System.Runtime.Caching**, in the **Microsoft.Practices.EnterpriseLibrary.Runtime.Caching** name-space. If a future version of Silverlight does introduce the **System.Runtime.Caching** to Silverlight, it will be relatively easy to swap out the Enterprise Library implementation with the Silverlight version.

Design of the Silverlight Caching Application Block

The following diagram describes the main classes that are used by the caching block.

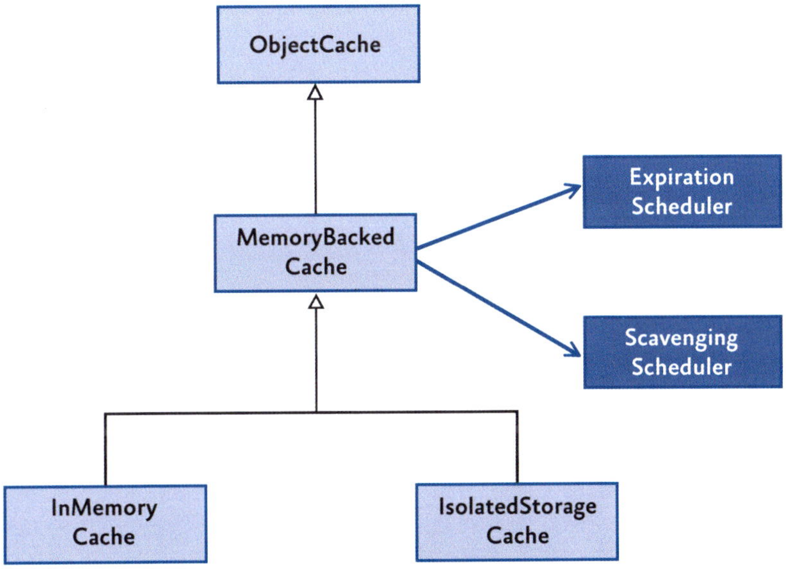

As you can see, there are two implementations of a cache, called the **InMemoryCache** and the **IsolatedStorageCache**. Ultimately, they both derive from an abstract base class called **ObjectCache**.

Object Cache and Memory-Backed Cache

The **ObjectCache** class is the abstract base class in **System.Runtime. Caching** that all cache implementations must derive from. It provides methods for adding, retrieving, and removing objects from a cache.

The **MemoryBackedCache** base class contains common functionality for both the **InMemoryCache** and the **IsolatedStorage Cache**. This base class also performs the expiration and scavenging logic based on notifications coming from two schedulers running on a background thread. The expiration scheduler will notify the cache if there are items that have expired. The scavenging scheduler will tell the cache to start removing items if the maximum size of the cache has been exceeded. For the **InMemoryCache**, the maximum size is determined by the number of items in the cache, whereas for the **IsolatedStorageCache**, the maximum size is determined by the space available on disk.

Cache Item Policy

When you add an item to the cache, you must also specify a cache item policy. This cache item policy allows you to control when an object will be removed from the cache. For example, you can specify a **SlidingExpiration** timespan for a cache item. This means that the item will be removed from the cache if it hasn't been accessed within the specified period. You can also specify an absolute DateTime to control when the item will expire.

The fact that you can specify expiration doesn't guarantee that the item is not removed before it has expired. For example, if the cache becomes too large, it can be removed by the scavenging logic.

> **Note:** *The expiration logic is performed on a background thread at a configurable interval. This means an item is not removed from the cache exactly when the item expires, but will be removed sometime after the item expires. Normally this doesn't cause issues, because most operations on the cache check if an item has expired. For example, if you test if an item is stored in the cache, the cache will return false if the item is still in the cache but has expired. When you check the actual contents of the cache using a debugger or query the number of items using the* **Count** *property, you'll see that it is still there until the expiration thread has removed it.*

The **CacheItemPolicy** class also allows you to specify two callbacks: the **RemovedCallback**, which gets called when the item gets removed from the cache, and **UpdatedCallback**, which is called when a cached value is updated.

> **Note:** *The* **UpdateCallback** *and* **RemovedCallback** *in the* **CacheItemPolicy** *class can be used both in the in-memory cache and the isolated storage cache. However, for the isolated storage cache it is not possible to serialize the callbacks. This means that the callbacks are lost when the application closes. The only way to restore these callbacks is to explicitly add the cached item again.*

Cache Item Priorities

The **CacheItemPriorities** are used by the scavenging logic to remove items from the cache when the maximum size of the cache is reached. The cache item policy allows you to specify a cache priority. **System.Runtime.Caching** defines two priorities: **NotRemovable** and **Default**.

The scavenging logic will first try to scavenge items that have **CacheItemPriority.Default**. However, if it then turns out that the cache is still too large, it will also remove items with **CacheItem Priority.NotRemovable**. So the term **NotRemovable** is a bit misleading, because any item can be removed from the cache if the cache becomes too full.

Cache Capabilities

The .NET Framework version of the object cache also provides a mechanism for describing the capabilities of a particular cache implementation. In theory, if you have several cache implementations at your disposal, you could use this information to select a particular cache implementation based on its capabilities. The isolated storage cache and in-memory cache describe their cache capabilities, but the Enterprise Library 5.0 Silverlight Integration Pack does not ship with any mechanism to select a caching implementation based on its capabilities.

Thread Safety

The Silverlight Caching Application Block is thread-safe. That means you can read and write from the cache from different threads without worrying that the cache becomes corrupt. In fact, the expiration and scavenging logic is executed on a different thread.

Because the scavenging logic and expiration logic executes on different threads, it is possible that the contents of the cache change between calls. Therefore you should try to interact with the cache using atomic operations. The following code shows an example of this:

C#

```
// Correct way to retrieve something from the cache
//(using atomic operation)
object item = cache["CacheItemKey"];
if (item == null)
{
   // Item was not stored in the cache
}
// Incorrect way to test if something is in the cache
//(using two operations)
object item = null;
if (cache.Contains("CacheItemKey"))
{
   // Item might have been added or removed between calls.
   // You can't assume that item is not null
   item = cache["CacheItemKey"];
}
```

When you query the contents of the cache, for example when you iterate over the entire contents of the cache, the cache will create a snapshot of itself at the moment you retrieved the iterator. That means you can safely iterate over the contents of the cache without the iterator breaking. However, while iterating, you will not see any changes that are made to the cache.

In-Memory Cache

The **InMemoryCache** allows you to create a cache that stores data in memory. This is very similar to the **MemoryCache** class from the **System.Runtime.Caching** namespace on the desktop version of the .NET Framework. The information stored in the in-memory cache is lost when the application closes.

Using the in-memory cache is similar to holding object references to retrieved data in code. Both these techniques will store the results in memory. But the in-memory cache allows you to specify when the data will expire and also allows you to refresh the data once it has expired. If you use a lot of cached data, the in-memory cache can also help you to manage the size of the cache.

Limiting the Size of the In-Memory Cache

With the in-memory cache, you can set limits on how large you'd like the cache to grow. You can set an upper boundary on the maximum number of cache entries. When the maximum number of items is reached, the scavenging logic will then start to trim the cache. You can also specify how aggressive the scavenging logic should be by specifying how many items should be left in the cache after the scavenging is complete.

It's not possible in Silverlight to determine how many bytes an object actually uses in memory, so it's not possible to limit the cache to actual bytes used. Memory consumption will be much higher if you cache large objects or lists of objects, compared to caching many very small objects.

Isolated Storage Cache

Sometimes, you'll want data that's stored in the cache to be persisted to the file system so that it's available after the application starts. If you have data that doesn't change often and that is needed during the startup of your application, then storing this data in isolated storage can improve the startup time of your application.

> **Note:** *The data stored in isolated storage is not secured. This means that end-users and system administrators can locate the data stored in isolated storage and either read or change this data.*

Limiting the Size of the Isolated Storage Cache

By default, Silverlight will limit the size of isolated storage to 1 Mb of data. You can increase the size of the isolated storage quota, either through explicit user consent or through a .NET Framework configuration policy.

For most applications, you can ask the user for a larger isolated storage quota, but there is no guarantee that the user will permit it. To help with that situation, the isolated storage cache gives you several settings that you can use to control the size of the cache.

When you create the cache, you'll have to set its maximum size. While this means that the cache will not grow larger than the maximum size, it does not guarantee that the cache can actually grow to this size. For example if you have a 1 Mb quota and 300 Kb is already used by other data, then the actually available size for the cache is 700 Kb.

You must also specify when you want the scavenging logic to start and how much space the scavenging logic should reclaim. You do this by specifying percentages of the actually available space. For example, if you have 1 Mb of available space and specify that scavenging should start at 80% and stop at 50%, then the scavenging logic will start at 800 Kb and will trim the cache back to 500 Kb.

> **Note:** *The scavenging percentages look at the available free space, or the maximum size you have specified, whichever is lower. So if you have specified a maximum cache size of 500 kb, and the isolated storage quota is set to 1 Mb, the percentages will work on the maximum size. If you have specified a maximum size of 2 Mb, but there is only 0.5 Mb of free space available, then 0.5Mb is used.*

Item Size on Disk

The isolated storage cache stores each cache item in a separate file in isolated storage. Because the way the file system handles small files, each cache item will consume at least 1 Kb of space. This means that with the default size of isolated storage, you can store a maximum of 1024 files in the cache.

It is recommended that you do not store small items in isolated storage. For example, if you wish to store several items in the cache, it's recommended you store them as a list of items, rather than storing each item using a separate value in the cache.

You should ensure that items you store in the cache are serializable. If you try to add an object to the isolated storage cache that is not serializable, the add operation will fail silently and the item will not be added to the cache.

Multiple instances of your application might be started up more often than you might think. If a user has opened your application and presses ctrl+n to create a new browser window, he will have two instances of your application running at the same time.

JSON Serializer

Cache items are serialized in JavaScript Object Notation (JSON) format using the **DataContractJsonSerializer**. This means that the classes you'd like to store in the cache should be serializable to JSON. In general, this includes most primitive types, most array and collection types, as well as complex types that use **DataContractAttribute** and **DataMemberAttribute**. Please refer to the chapter on Stand Alone JSON Serialization on MSDN® for more information.

Multiple Instances of the Isolated Storage Cache Running at the Same Time

The space used in isolated storage is shared by all instances of your application. This causes problems for the isolated storage cache, because this would cause multiple instances of your application to read from and write to the same files. Synchronizing between multiple instances of the application would require complicated locking and synchronization logic, which would also have a negative impact on the performance of the cache.

To overcome this issue, only the first instance of your application will be given read and write access to isolated storage. For any subsequent instances that create an isolated storage cache, that cache will be automatically populated with all the data that was already stored in isolated storage at that point in time, but after the creation it will behave like the in-memory cache.

The advantage of this approach is that multiple instances of your application have the ability to cache data. If you populate isolated storage with data that is used often and does not expire easily, then all instances of your application should have roughly the same performance.

When the isolated storage cache is locked and starts to behave like an in-memory cache, it is no longer possible to detect how much memory the cache consumes. This means that the isolated storage cache cannot use the configured scavenging percentages because they work on the actual size it would have consumed on disk. To prevent this cache from growing beyond control, a limit will be placed on the maximum number of items the cache can hold. This number is determined by the maximum number of items that were already stored in isolated storage. To prevent the cache from becoming unusably small, this number is set to a minimum of 20 items.

For example: The first instance of the application has started and added 40 items to the isolated storage cache. Then the second instance of the application starts. It also creates the isolated storage cache, which will be populated with the 40 items that were added by the first instance. From then on, the second instance will work as the in-memory cache. If you then add an additional 10 items to the second cache, it will start scavenging and remove the 10 oldest items, to make room for the newly added 10 items.

Implementing Caching in the Stock Trader: The Position Module

In this chapter, we'll discuss how caching was implemented in the Stock Trader V2 application. To improve the startup performance, the position module stores the user's position and their market history in isolated storage. At the end of this chapter, we'll discuss how the news module uses interception to transparently cache retrieved news articles to the in-memory cache.

When we started to build the Stock Trader V2, we knew that performance was a very important aspect of the application. However, it's difficult and often not very effective to optimize the performance of your application much beforehand. For the Stock Trader V2, we waited to make performance optimizations until we had actual performance and usage information.

First, we focused on implementing the functionality of the Stock Trader V2. Using an early prototype of the application, we conducted usability studies and early performance tests. By looking closely at the usage scenarios of the application and comparing those to the performance tests, we identified the areas where caching and other optimizations would provide the most benefit. The result was an application that performed really well.

Scenario

In the position summary view, the users get an overview of the stocks they own, their current market value, and their cost basis. When the user selects a particular stock, a graph displays the market history in a trend line and related news articles are displayed.

From the usability studies we performed while developing the Stock Trader V2, we found out that most of our users start the application many times a day, to check up on the current value of their stocks. For these users, it was very important for the application to start as quickly as possible.

The second most used feature of the application is the trend line. As soon as they started up the application and noticed a change in the price of their stock, they wanted to see if this change is part of a trend. Users reported that it was annoying to have to wait until the trend line was loaded. They also reported that when they quickly change between stocks, the wrong trend line would sometimes be displayed.

Lastly, a place where we found a performance bottleneck was in the news module. Whenever you select a stock, we also download and display the latest news messages. We found that most users will change the selected stock very often. This would cause way too much traffic when an important event in the stock market caused there to be many news articles.

Design of the Position Module

The following diagram shows a schematic design of the position module.

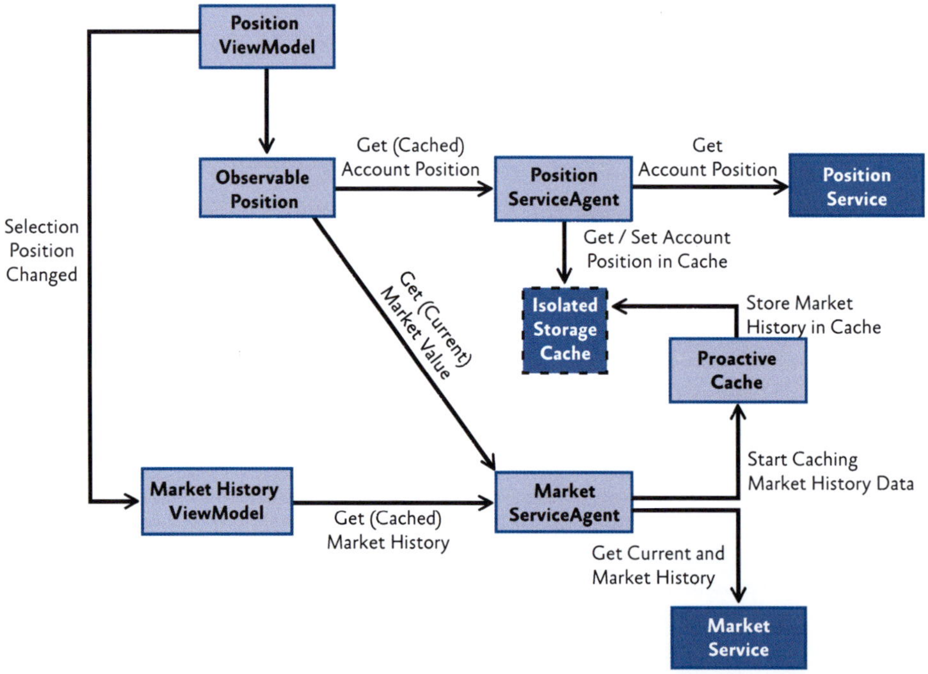

The position view, with its associated position view model, displays a list of the current account positions, with their current market values. The current account position information from the Position Service Agent and the current market value information from the Market Service Agent is aggregated and continuously updated in the **ObservablePosition** class.

The two service agents isolate the view models from all the details about how the particular information is being retrieved.

> While building the Stock Trader V2, we isolated all interactions with web services in special service agent classes. This had the advantage of making our user interface logic much easier to test using unit tests. But it also provided us an isolated location for applying caching. We only wanted to apply caching to places where it really improved the performance and user experience. Because we had isolated the logic for accessing web services in separate service agents, we were able to apply caching to the application without having to change the UI logic.

Caching Values to Isolated Storage

The position service agent will cache requests for the current account position to isolated storage. Once this information is cached, the application would start much faster. Since the current account position data also contains the market value on the last closing day, we decided that we wanted to invalidate the cache at the start of the next trading day.

Proactive Caching

When the user selects a particular stock, the position view model will notify the market history view model to display the information for that particular stock. The market service agent will then asynchronously retrieve the information from the service. The market history only needs to be updated once a day, so by storing this information in isolated storage, the users can view the market history much faster after starting the application.

> Our users reported a nasty problem with the market history trend line. When they quickly changed between different stocks, occasionally they would see the wrong trend line. After some examination, we found that this was caused by the market service agent. This service agent retrieves the market history asynchronously. When you select stock 1, it will start downloading the market history for stock 1. If you then quickly select stock 2, it will also start downloading market history for stock 2. However, if the market history for stock 1 has completed its download, it will begin to display. Only when stock 2 has completed its download will the market history of stock 2 be displayed.

To solve this problem, but also to improve the performance of the trend line, a proactive caching implementation was developed. This means that the market history information would be downloaded and added to the cache on a background thread. When the user selects a particular stock and the market history is not yet downloaded for this stock, the market history for this stock will be downloaded as soon as possible.

The following diagram shows how the proactive cache works:

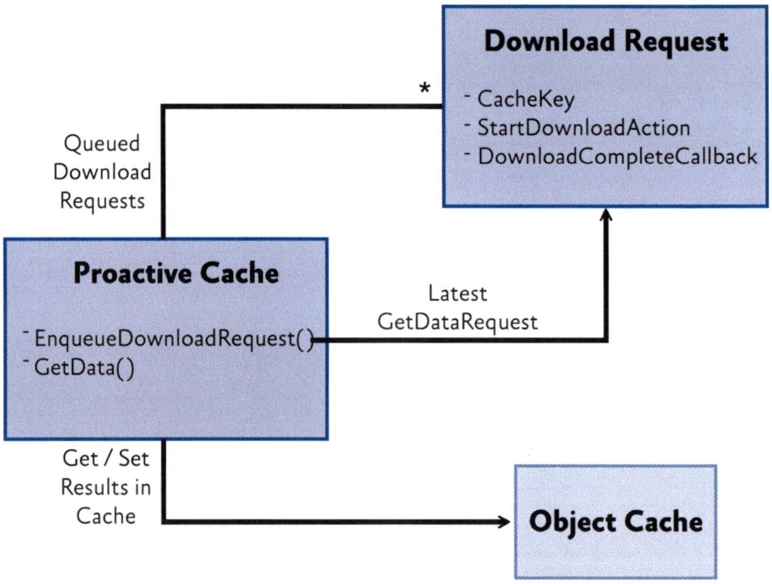

At the core of the proactive cache is the **DownloadRequest** class. This download request has two actions: The **StartDownloadAction**, which will start the download if you call it. It can also have a **Down loadCompleteCallback**, which will be called when the download has completed. This callback is typically used by the user interface to detect that the download has been completed and that the data can be displayed.

The download request also has a **CacheKey** property. Before starting the download, the **ProactiveCache** will check if an item with that **CacheKey** is already present in the cache. If so, it will return the cached item directly. If the item with that cache key is not found, then the asynchronous download is started. After the download has completed, it will add the downloaded value to the cache under the specified **CacheKey** so any subsequent calls don't need to download again.

The proactive cache has two public methods: **EnqueueDown loadRequest** and **GetData**. The **EnqueueDownloadRequest** method will add download requests to the download queue, which will be processed by a background thread in the order received. The **Get Data** method will add a download request that will be processed as soon as possible. This method will remember the last request that was added by the **GetData** method. Only for the last request will the **DownloadCompleteCallback** be called. This means if the user very quickly selects three different stocks, then the trend line view will only be updated for the last selected stock. The market history for other stocks will eventually be downloaded in the background, so if the user selects them later, the item can be returned from the cache very quickly.

> Proactive caching is a great technique for preemptively retrieving information that the user is very likely to use. When used in the correct places, this technique can improve the user experience. However, you should keep in mind that downloading information in the background can cause user-initiated downloads to be slower. It is also possible that you'll download more information than the user actually needs. This will increase the pressure on your back-end servers, so you should only use proactive caching where it really makes sense.

Asking the User for More Space in Isolated Storage

When the Stock Trader starts, one of the first actions it performs is to check the isolated storage quota. If the isolated storage quota is set below 10 Mb, it will ask the user to increase it.

In Stock Trader V2, the logic to increase the size of the isolated storage quota is found in the **IsolatedStorage** service class. This class has methods to query the isolated storage size, and to ask the user for a larger quota. Asking the user for a larger quota must always occur as the result of a button click. In other words, you will need to display a custom message asking the user for additional quota. This is the dialog box that the Stock Trader shows to ask for an increased quota:

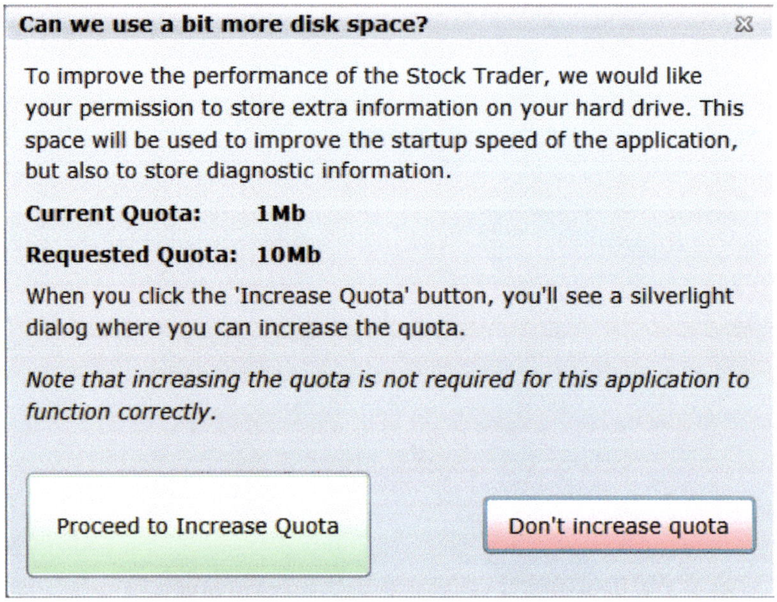

If the user agrees, then you can call Silverlight to try to increase the quota. This Silverlight dialog box will ask the user to increase the quota, and if the user agrees, then the quota is increased.

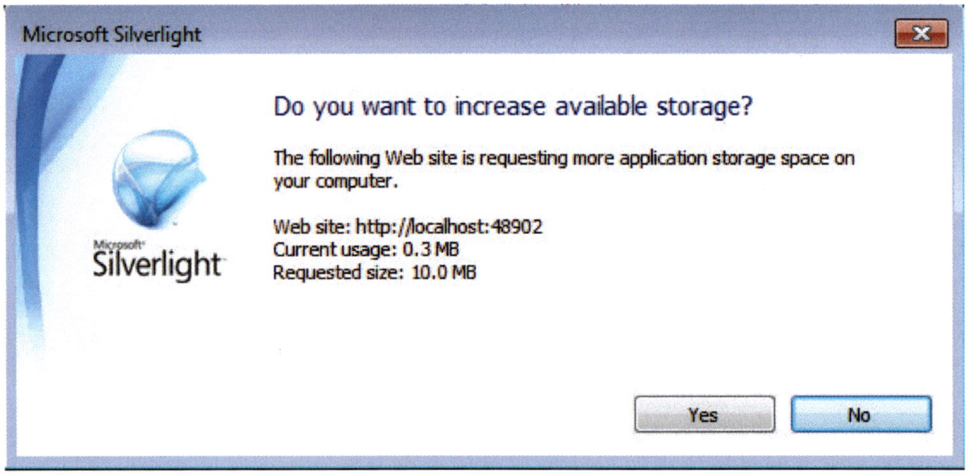

The following code example shows how you can increase the size of the isolated storage quota:

```
C#
IsolatedStorageFile isolatedStorageFile;
// Use the Quota property to find out what the current quota is.
long quotaInBytes = isolatedStorageFile.Quota;
// Use the IncreaseQuotaTo method to ask the user to
//increase the quota.
long newQuotaSize = 10*1024*1024; // 10 Mb
bool increased = isolatedStorageFile.
   IncreaseQuotaTo(newQuotaSize);
```

Implementing Caching in Your Application

In this chapter, we'll discuss how you can apply the Silverlight Caching Application Block to your application to cache values. First, you will need to configure which cache you would like to use. Then you can retrieve the object cache from the container and start to interact with it.

Configuring the Caching Application Block

Now let us look at how you can add either the in-memory cache or the isolated storage cache to your application through XAML configuration. If you want, you can use both caches at the same time, as long as they each have a unique name.

Adding an In-Memory Cache Using Configuration

To add the in-memory cache to your application using a configuration file, you can follow these steps:

1. Start the Enterprise Library configuration console
2. Add the Caching Application Block for Silverlight:

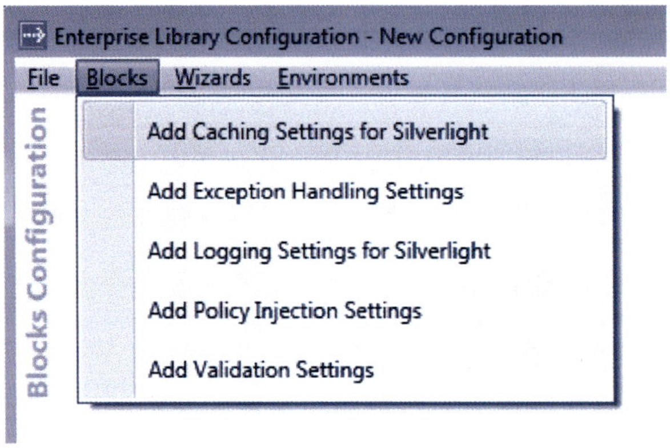

3. The **In-Memory Cache** is added by default. Configure the **In-Memory Cache** to fit your scenario. The name that you specify here is also the name that you should use when you wish to retrieve this cache from the container.

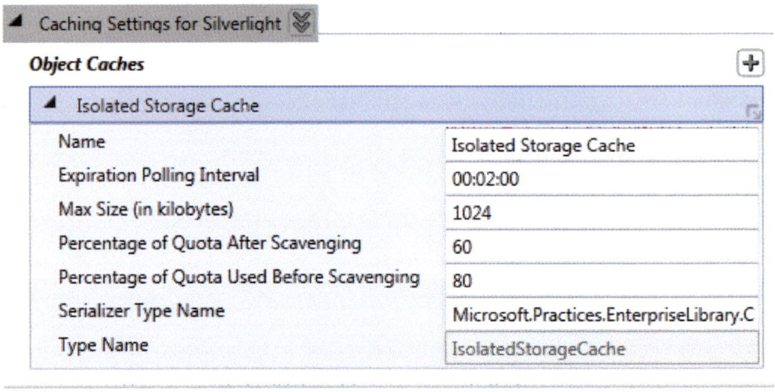

4. Select **Wizards -> Export to XAML**. Browse to your Silverlight project and save it under the name Configuration.xaml.
   Add the configuration.xaml file to your Silverlight project and set the build action to **Content** in the file properties.

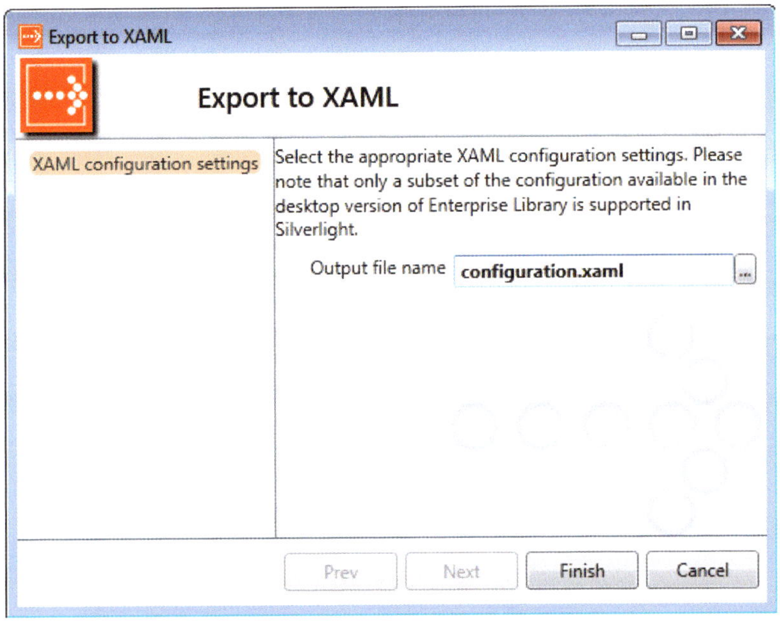

Adding an Isolated Storage Cache Using Configuration

1. Start the Enterprise Library configuration console
2. Add the Caching Application Block for Silverlight:

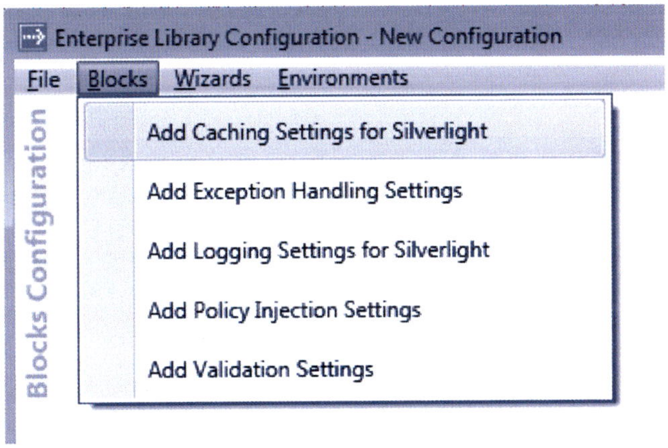

3. Remove the **In-Memory Cache,** which is added by default, by right clicking the **In-Memory Cache** and selecting **Delete In-Memory Cache**.

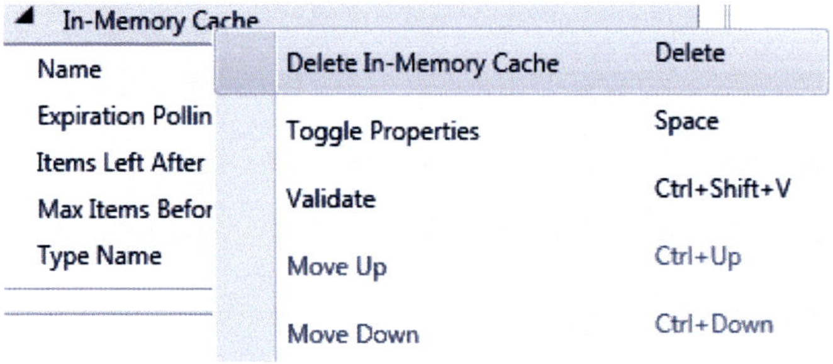

4. Add the Isolated Storage Cache by clicking the plus (+) sign and selecting **the Isolated Storage Cache**:

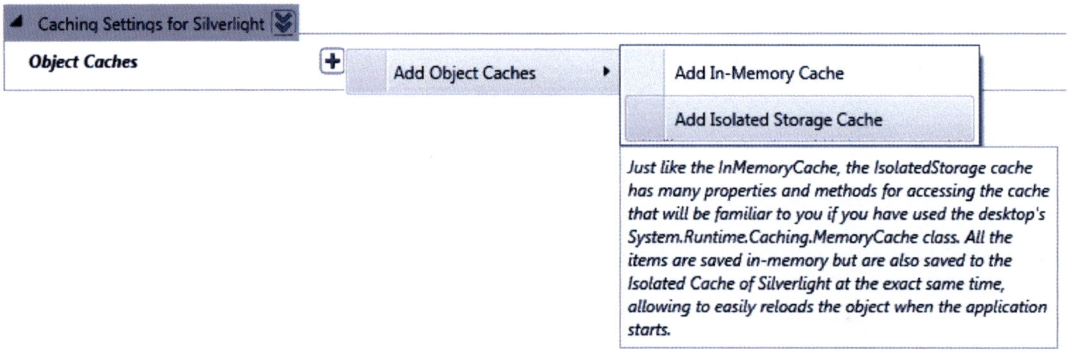

5. Configure the Isolated Storage Cache to fit your scenario. The name of the Isolated Storage Cache is used for retrieving the correct cache implementation from the container.

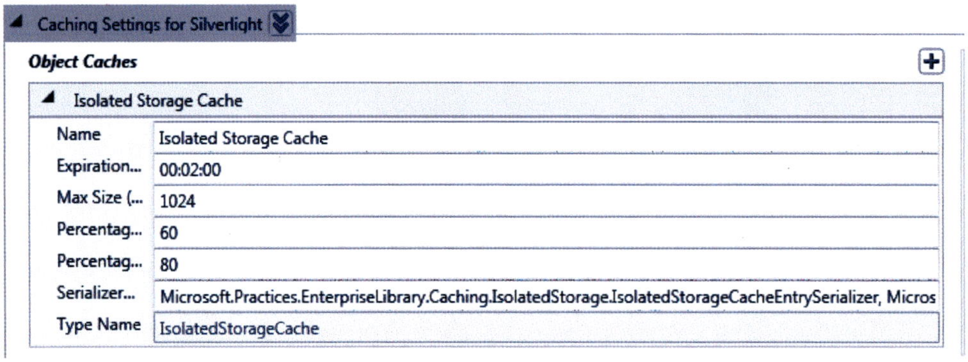

6. Select **Wizards -> Export to XAML**. Browse to your Silverlight project and save it under the name Configuration.xaml.

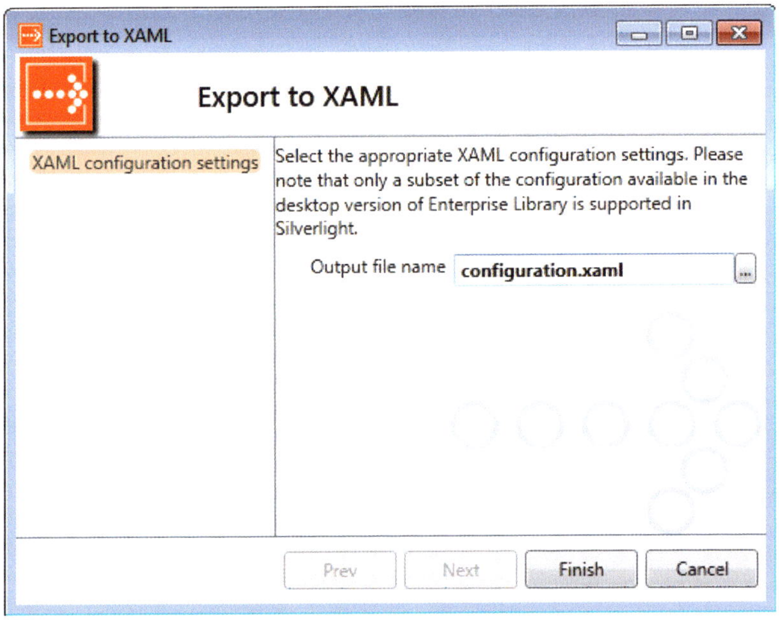

7. Add the Configuration.xaml file to your Silverlight project and set the build action to **Content** in the file properties.

Using the Object Cache

Once you have configured either the in-memory cache or the isolated storage cache, you can use it from the container. It's recommended that you use the cache through the **ObjectCache** abstract base class, because this allows you to change the physical implementation without having to change your code.

Retrieve the Cache from the Container

The following example demonstrates how you can retrieve the object cache from the container:

```
C#
ObjectCache cache = EnterpriseLibraryContainer.Current.
GetInstance<ObjectCache>(
    «CacheName»);
```

You should set the string value "CacheName" to the name you gave your cache when you configured it.

Adding Items to the Cache

The following code sample shows how to add an item to the cache. It also shows several options for using cache item policies.

```
C#
object valueToSet = new Object(); // Set this to the value to
                                  // add to the cache
// Example of an absolute expiration policy.
// Item will expire in 2 minutes, regardless if it is
//accessed or not
var absoluteExpirationPolicy = new CacheItemPolicy{
   AbsoluteExpiration = DateTime.Now.AddMinutes(2)};
// Example of a sliding expiration cache item policy item
// will expire if not accessed for more than 2 minutes
var slidingExpirationPolicy = new CacheItemPolicy{
   SlidingExpiration = new TimeSpan(0, 0, 2, 0)};
// Example of a policy that defines that item will not expire.
// It will only be removed if the cache becomes full.
var notExpiringPolicy = new CacheItemPolicy();
// The add method will only add the item if it doesn't
// already exist
bool added = cache.Add("CacheKey", valueToSet,
   notExpiringPolicy);
// The set method will add or overwrite an item.
cache.Set("CacheKey", valueToSet, notExpiringPolicy);
// When setting a cache item using the indexer, it will always
//be set to infinite expiration
cache["CacheKey"] = valueToSet;
// The AddOrGetExisting method will try to add it if it doesn't
// already exist. If it has added it, then it will return the
//added item, else it will return the already cached item.
var cachedValueOrValueToSet = cache.AddOrGetExisting("CacheKey",
   valueToSet, notExpiringPolicy);
```

Testing If an Item Is in the Cache

The following code example shows the recommended way to test if an item is already in the cache:

```
C#
var cachedItem = cache["Cachekey"];
if (cachedItem == null)
{
   // Item was not cached
}
else
```

```
{
   // Item is cached
}
```

```
C#
// Specify that the ItemRemoved method should be called when
// the item gets removed
var policyWithRemovedCallBack = new CacheItemPolicy {
   RemovedCallback = ItemRemoved};
cache.Set("CacheKey", valueToSet, notExpiringPolicy);
cache.Remove("CacheKey");
private void ItemRemoved(CacheEntryRemovedArguments arguments)
{
   // The arguments object contains information about the
   //removed item such as:
var key = arguments.CacheItem.Key;
var removedObject = arguments.CacheItem.Key;
var removedReason = arguments.RemovedReason;
}
```

While you can use the **Contains** method to test if something is in the cache, this method is not recommended. Due to the multi-threaded nature of the cache, it is possible that the cached item is removed directly after calling the **Contains** method.

Using the Removed Callback

You can use the **RemovedCallback** to detect that an item is removed from the cache. The following example demonstrates how to use this:

```
C#
// Specify that the ItemRemoved method should be called when
// the item gets removed
var policyWithRemovedCallBack = new CacheItemPolicy {
   RemovedCallback = ItemRemoved};
cache.Set("CacheKey", valueToSet, notExpiringPolicy);
cache.Remove("CacheKey");
private void ItemRemoved(CacheEntryRemovedArguments arguments)
{
   // The arguments object contains information about the
   //removed item such as:
var key = arguments.CacheItem.Key;
var removedObject = arguments.CacheItem.Key;
var removedReason = arguments.RemovedReason;
}
```

These callbacks do not get serialized to isolated storage. When the application is closed, these callbacks are lost and will not be executed again when the application starts and these items are evicted from the isolated storage cache.

Using the Updated Callback

You can use the **UpdatedCallback** to detect that a cached item or its Cache Item Policy has changed. The following code sample shows how this works:

C#

```
var policyWithUpdateCallBack = new CacheItemPolicy {
   UpdateCallback = ItemUpdated };
cache.Set("CacheKey", valueToSet, notExpiringPolicy);
cache.Set("CacheKey", changedValue, notExpiringPolicy);
private void ItemUpdated(CacheEntryUpdateArguments arguments)
{
   var key = arguments.Key;
   // The updated Cache Item allows you to make changes
   //to the cached item
   var updatedCacheItem = arguments.UpdatedCacheItem;
   // The updated cache item policy allows you to make
   //changes to the cache item policy
   var updatedCacheItemPolicy = arguments.UpdatedCacheItemPolicy;
   var removedReason = arguments.RemovedReason;
}
```

8 Interception

Since the invention of computers, programming languages have been evolving, and each improvement has increased the level of abstraction.

While modern programming languages such as C# 4.0 are much more powerful than their predecessors, you will find that programming is still as much about addressing technical programming issues as it is about addressing business requirements. Orthogonal issues such as authorization, exception handling, caching, and diagnostic logging often require a lot of code, which is frequently mixed with code that addresses actual business requirements.

In this chapter, we'll discuss an advanced programming technique called interception. When used properly, interception can help you to reduce the amount of code you'll have to write to solve technical programming issues and will allow you to focus on addressing actual business requirements.

Unity Interception has been available for the Microsoft® .NET Framework as an extension for Unity container for several years now. Now it is available for Microsoft Silverlight as well. Here, we'll discuss what interception is, why you might want to use it, and how you can apply Unity Interception and policy injection to Silverlight applications.

What Is Interception?

Interception is an advanced programming technique that allows you to intercept calls to an object so that you can add additional logic before or after the call. This interception process should be virtually transparent, so both the calling object and the target object can be oblivious to the interception process.

The following diagram shows how instance interception typically works:

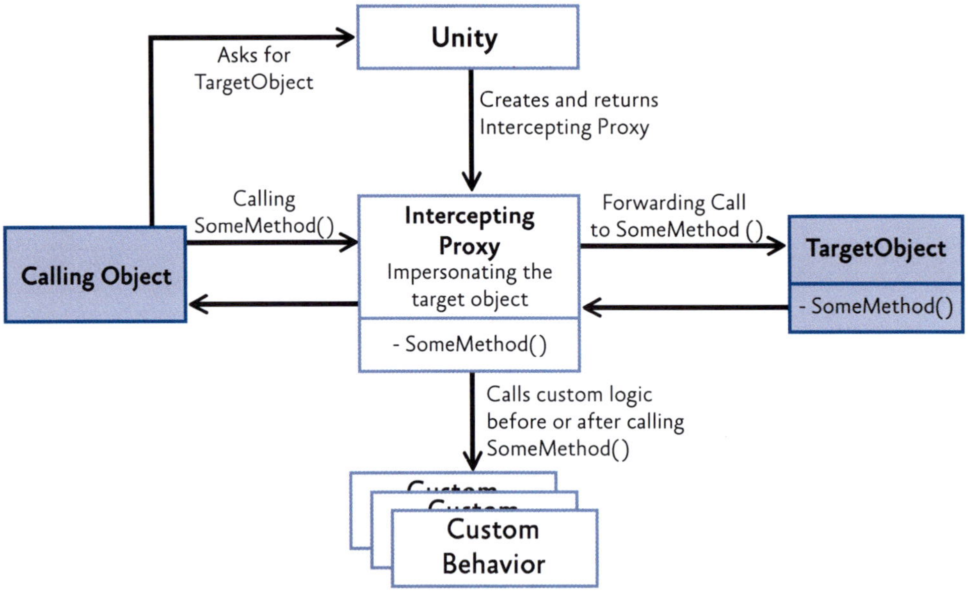

The calling object will typically ask the Unity dependency container for a target object. However, it will not receive a real **Target Object**, but rather an intercepting proxy that is impersonating the real **TargetObject**, for example, by implementing the same interface. This way the calling object does not know if it calls the real target object or the proxy. The **TargetObject** also doesn't know that the calls have been intercepted. The fact that both classes don't know this about the interception process allows you to add custom logic before or after a member call, without having to change either the calling object or the target object.

Interception is quite similar to using the decorator pattern. With a decorator you'll typically have to program a custom decorator class for each target object. However, Unity will dynamically generate the interceptor for you, so all you have to do is implement the custom behaviors.

Interception to Handle Cross Cutting Concerns

Interception is very well suited to implementing cross cutting concerns. A cross cutting concern is an aspect of programming that typically affects many parts of an application. One example of a cross cutting concern is handling exceptions at certain layers.

Exception Handling as a Cross Cutting Concern

If you implement exception handling directly in code, your classes will likely be mingled with logging statements. For example, in your user interface you will probably want to catch all unhandled exceptions, show a friendly error message to the user, and log the exception information. At service boundaries, where your application makes requests to external web services, you might want to translate any service exceptions into more friendly exception types that your application can more easily handle.

For relatively small applications it is usually not a problem to spread exception handling logic around the application. However, for larger enterprise applications this can cause serious problems.

- While cross cutting concerns, such as exception handling, are very important to your application, they usually don't add any direct business value. If you have implemented exceptions only for exceptional cases, your application will likely function fine without exception logic—as long as nothing goes wrong, of course. But adding these exception handling statements to your business logic means that your code will be cluttered with try-catch statements, which makes it harder to read and harder to maintain.
- Implementing cross cutting concerns such as exception handling directly in your code can also cause other issues for large, enterprise-scale applications. For example, not every developer will use the same level of diligence when applying exception handling. It's then likely that inconsistencies will arise in the exception handling code, with nasty stability problems resulting. If these inconsistencies were to occur not just in exception handling but also in other cross cutting concerns, such as authentication, the result could be major security problems.
- The fact that these cross cutting concerns are spread throughout your application can make them very difficult to change. For example, if you have applied exception handling very diligently and most of your classes are handling exceptions, then it becomes a lot harder to change to a different exception handling strategy, because that would mean changing all of the

classes that perform exception handling. Also, if you were to decide that you're not handling exceptions at the right locations in your application, then you would have to manually change all these classes to alter the locations where you are handling exceptions.

- If you would have implemented exception handling using interception, then you would have been able to isolate your exception handling logic in specific interception behaviors and dynamically apply them where you need them.

Other Examples of Cross Cutting Concerns

Exception handling is only a single example of a cross cutting concern. Typical enterprise applications have dozens, such as:

- Logging and tracing information as it flows through your application.
- Executing calls to back-end web services under different security credentials.
- Authenticating calls to business components, so only authorized users can access them.
- Implementing validation logic on methods of your business components.
- Caching data retrieved from a web service or database.
- Again, it's certainly possible to address all these cross cutting concerns directly in your code, but this will pollute your business logic with code that's not directly solving the business problem at hand.

Benefits of Applying Interception to Cross Cutting Concerns

There are several major benefits to using interception to solve certain cross cutting concerns. For example, if you want to authenticate all calls to your business components, you would typically need a lot less code to implement such a cross cutting concern using interception. Instead of having to add authentication code to each of your business components, you can write a single interception behavior that handles authentication and apply it to all your business components.

Another advantage is that your interception behaviors can often be easily reused in other places. In the case of the authentication behavior, if you later decide that calls to other components should also be authenticated, you can simply apply that interception behavior to those components as well.

The fact that you can easily apply a behavior to several classes also means that you'll get more consistency in your cross cutting concern. It's much easier to ensure that all calls to your business components are authenticated if you don't have to add authenticating code into each method. When using the interception behavior, the authentication is applied automatically.

Lastly, implementing authentication using interception also makes your application more maintainable. That's because the business logic is easier to read as it is not polluted with authentication statements; plus, you've also centralized authentication in a single location. This makes it much easier to change the way authentication is handled.

Unity Interception for Silverlight

With the introduction of Enterprise Library 5.0 Silverlight Integration Pack, Unity Interception has now been almost fully ported to Silverlight. In this chapter, we'll discuss the features that Unity Interception provides, which scenarios it supports, and how to apply it.

Unity Interception

Unity provides a built-in mechanism for performing interception. When you register a type with Unity, you also have the option to register an interceptor, which describes how the target object should be intercepted, and one or more behaviors, which control what actions to perform when member calls are intercepted.

Basically, there are two methods for implementing interception: instance interception and type interception. Both have specific advantages and disadvantages that you'll need to be aware of.

Instance Interception

One method of interception is called instance interception, which essentially wraps a target object with a proxy that impersonates the target object. If a member call can be intercepted, it will be routed through the proxy before it reaches the target object. This is where the proxy can apply interception to those member calls. The following diagram explains in more detail how this interception process works.

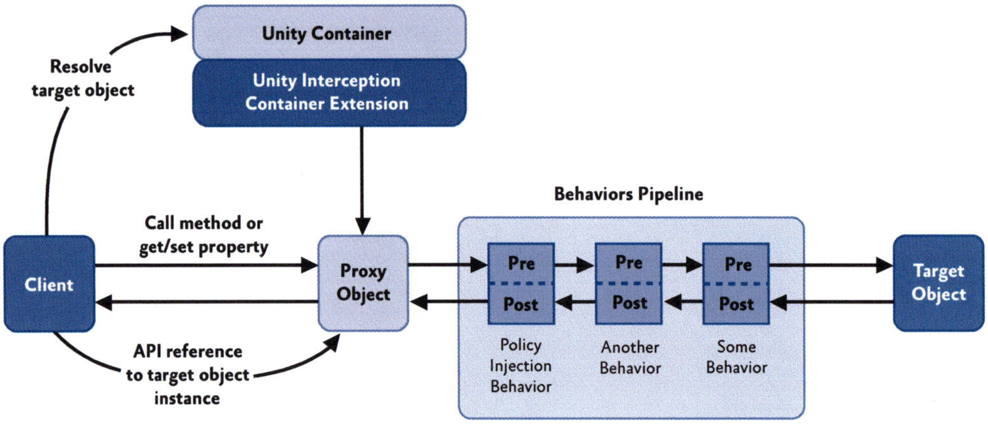

The typical occasion for using instance interception is when the client uses Unity to resolve a new instance of the target object. Unity will create both the target object and a proxy object that wraps the target object and returns the proxy to the client, rather than the target object. Any calls by the client to the proxy object are fed through a pipeline of behaviors until they reach the target object's implementation of the invoked method. These behaviors allow you to apply custom logic before or after the member call. Please refer to the section on **Interception Behaviors** for more information on this pipeline of behaviors.

In Silverlight, the only way the proxy can impersonate the target object is through interfaces. The proxy will then implement the one or more interfaces as the target object (you can choose). In Unity, this is accomplished using the **InterfaceInterceptor** class.

Note: On the desktop versions of the .NET Framework and Enterprise Library, it is also possible for the proxy to use .NET remoting to impersonate the target object. In Unity, this is implemented using the **TransparentProxyInterceptor**.

InterfaceInterceptor

If you use the **InterfaceInterceptor**, the interceptor will automatically create a proxy object for you that implements the requested interface. The advantage of **InterfaceInterceptor** is the enhanced performance of the interception process, but it has as a down side in that you can only intercept the members of the requested interface.

TransparentProxyInterceptor

The **TransparentProxyInterceptor** uses .NET remoting to create a proxy. In order to be able to do that, the target object needs to derive from **MarshalByRefObject**. This method has as an advantage that you can intercept any member call to the target object, but these calls will be a lot slower than regular member calls. Since .NET remoting is not available in Silverlight, this option is only available in the desktop version of the .NET Framework and Enterprise Library 5.0.

> **Note:** *The* **TransparentProxyInterceptor** *does not work properly if you hide inherited members using the* **new** *keyword. If you use the* **new** *keyword for a member, effectively you'll create two members with exactly the same name and signature. The remoting proxy then has no way to distinguish between them and will throw an* **Invalid OperationException**.
>
> *Using the* **new** *keyword is generally not recommended because it can cause undesired side effects when you use polymorphism, but if you find that you really need to do so, you should be aware that you cannot use the* **TransparentProxyInterceptor**.

Advantages and Limitations of Instance Interception

A major advantage of using instance interception, compared to type interception, is that it is possible to apply interception to existing object instances. While it is possible to resolve newly created intercepted objects from the Unity container, it is also possible to apply this type of interception to already created instances. This is especially useful if you are not able to construct the target objects yourself, for example, but must rely on a factory method to instantiate the objects.

A limitation of instance interception is the fact that only external calls to the object, which go through the proxy, can be intercepted. So if the target object calls a method on itself, the call will not be intercepted. That sort of interception can only be provided by type interception.

For an object to be interceptable using instance interception, it either has to implement an interface and use the **InterfaceInterceptor,** or derive from **MarshalByRefObject** and use the **TransparentProxy Interceptor**.

Type Interception

Type interception does not create a separate proxy, but creates a derived object from the target object. Type interception can intercept method calls by overriding virtual members. The overridden virtual member can then intercept the member call and apply custom interception logic before it calls the original base member. The following diagram shows type interception in more detail:

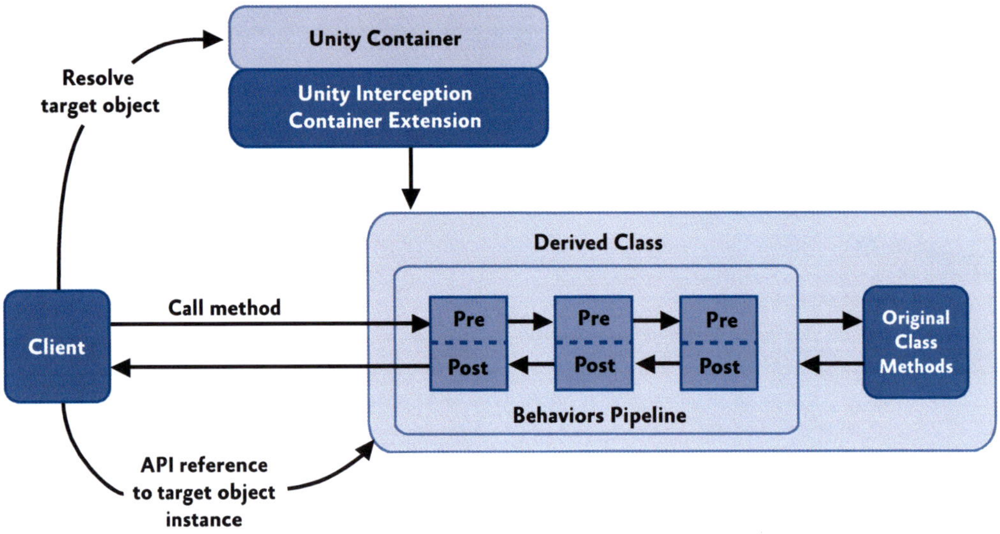

When the application resolves the required type through the Unity container, the Unity interception container extension creates the new derived type and passes it, rather than the resolved type, back to the caller. Because the type passed to the caller derives from the original class, it can be used in the same way as the original class. The client simply calls the target object, and the derived class will pass the call through the behaviors in the pipeline just as is done when using instance interception.

Because type interception uses inheritance, this method of interception can only intercept members that have been marked as virtual. The interception mechanism will dynamically create a base class for the target object and override the methods that have been marked as virtual.

Advantages and Disadvantages of Type Interception

Type interception uses inheritance to merge both the proxy and the target object into a single object instance. The advantage of this approach, compared to instance interception, is that internal member calls will also be intercepted.

Type interception is a bit more invasive than instance interception. You can only use type interception when you create a new instance of a type. Contrary to instance interception, it's not possible to use type interception on a previously created object instance. So if you cannot create the target object yourself, then you cannot apply type interception.

Another limitation of type interception is that you can only intercept members that you have explicitly marked as virtual. This means that you may need to modify the target class somewhat to apply interception.

Interception Behaviors

You can add custom logic to an intercepted member call by adding interception behaviors to the interceptor. An interception behavior is a class that implements the **IInterceptionBehavior** interface. You can register more than one behavior to an interceptor. These behaviors are chained together following the Chain of Responsibility pattern. When a call is intercepted, each behavior object executes its own logic and will then call the next behavior. The last behavior will then invoke the actual object. The result of the call (for example a return value or an exception) will be passed back through the chain of behavior objects until it reaches the calling object again.

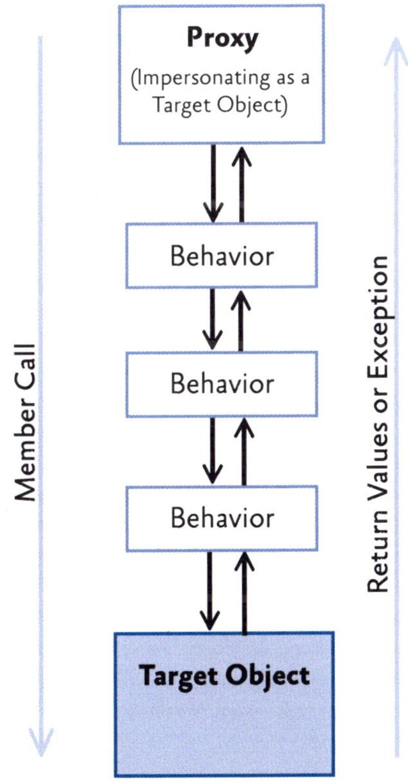

If you don't wish the target object (or the next behavior) to be called, you can prevent that from happening by not invoking the next behavior in the chain. There are several scenarios in which this is useful. For example, when you want to validate the input of a method call, and the input is not valid, you can prevent the target method from being called with invalid data and returning an exception. Another example would be when you'd like to cache the results of a method call. The caching behavior can check if the results of this particular method call are already present in the cache. If so, it can just return the cached result and avoid the target method being called.

For more information on implementing a custom behavior, please refer to the section called **Creating a Custom Interception Behavior**.

Interception and Policy Injection

While interception is a very powerful technique, it is also a bit coarse-grained. Once you set up an object for interception, each interception behavior in the pipeline will receive all interceptable member calls. Each of these behaviors must be able to receive all these member calls. If certain behaviors are designed to only work on specific members, then the behaviors themselves need to know how to target these members and leave other member calls unaffected.

Policy Injection gives you much more fine-grained control over the interception process. On one hand, it allows you to target very specific members for interception. But a policy does not need to be bound to a specific class. It can also be used to target whole families of classes (for example all service agents or view models) with a single policy. The following diagram shows how policy injection behavior works when doing instance interception:

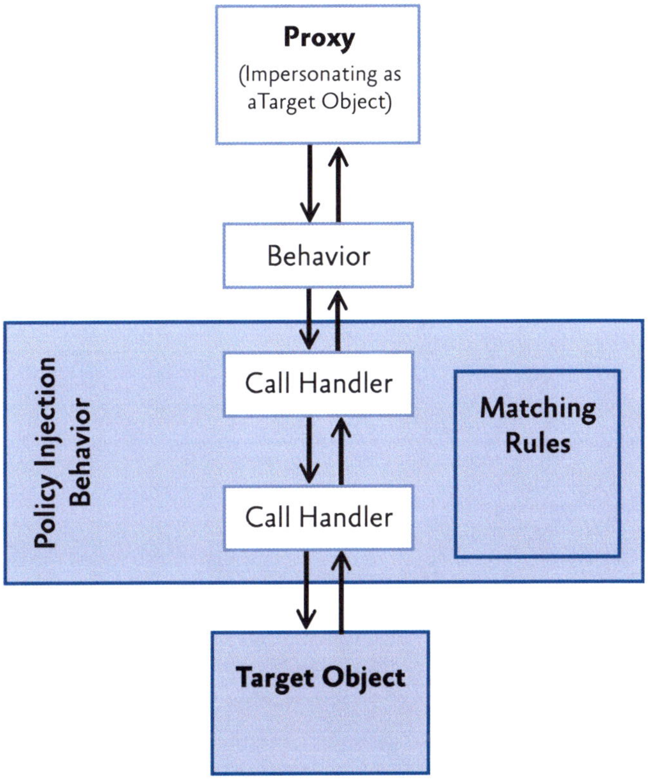

An interception policy consists of two parts: Matching rules and call handlers. The matching rules define which call handlers should be applied in the call to specific members of the classes being intercepted, while the call handlers do the actual interception.

As you can see, policy injection is defined as an interception behavior. Therefore, in order to use policy injection on a type, you must set up either type or instance interception and have the policy injection behavior applied.

Matching Rules

Matching rules allow you specify which classes and/or members should be intercepted. You must specify at least one matching rule. All the matching rules in the policy must be satisfied in order for the call handlers to be executed.

While it is possible to create your own matching rules, there are several matching rules that ship with Unity out of the box.

Matching rule	Description
Assembly Matching Rule	Selects classes in a specified assembly.
Custom Attribute Matching Rule	Selects classes or members that have an arbitrary attribute applied.
Member Name Matching Rule	Selects class members based on the member name.
Method Signature Matching Rule	Selects class methods that have a specific signature.
Namespace Matching Rule	Selects classes based on the namespace name.
Parameter Type Matching Rule	Selects class members based on the type name of a parameter for a member of the target object.
Property Matching Rule	Selects class properties by name and accessor type.
Return Type Matching Rule	Selects class members that return an object of the specified type.
Tag Attribute Matching Rule	Selects class members that carry a **Tag** attribute with the specified name.
Type Matching Rule	Selects classes that are of a specified type.

By using these matching rules, you can be very flexible in determining which policies should target which classes and members:

For example, you may want to create a very specific policy that only targets a couple of methods on a class. In that case, you could specify the type matching rule to target the specific type and a member name matching rule or the method signature matching rule. The member name matching rule allows you to pass a wildcard to be a bit more flexible in specifying which names you'd like to target.

In other situations you might want to create a very flexible policy that targets an entire family of classes. For example, if you have all your service agents in a separate namespace, then you can use the namespace matching rule to set up a policy that targets all interceptable members of all the classes in that namespace.

Please refer to the **Unity reference documentation** on MSDN for more information about the specific matching rules.

Call Handlers

A call handler is a class that implements the **ICallHandler** interface. This interface has an **Invoke** method that is called when a member is intercepted, just as with the **IInterceptionBehavior** interface.

Because you have more flexibility in how you would use interception, a call handler can either be written very specifically for a particular method, or very generally, to handle any member.

Built-in Call Handlers

Enterprise Library ships with a couple of call handlers out of the box. These built-in call handlers can help you perform the following actions:

- Logging method invocation and property access
- Handling exceptions in a structured manner
- Validating parameter values
- Authorizing method and property requests
- Measuring target method performance

9 Interception in the Stock Trader: The News Module

In this chapter, we'll discuss how we've implemented interception in the Stock Trader V2 reference implementation.

Caching through Interception

To improve the performance of the news module, the news articles have to be cached. The News Module demonstrates how you can implement caching to asynchronous services using interception. The following diagram shows how the interception works.

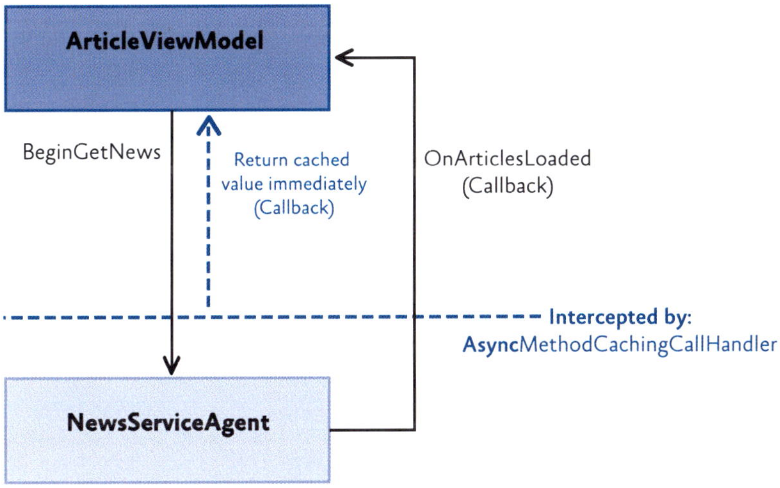

The ArticleViewModel initiates an asynchronous download by calling the BeginGetNews method on the NewsServiceAgent. When the download completes, the NewsServiceAgent will execute a callback that triggers the OnArticlesLoaded method on the ArticleViewModel.

The AsyncMethodCachingCallHandler intercepts the call to the BeginGetNews method. This call handler does require a specific method signature to work. The first parameter needs to identify the values to be cached; for example, the stock symbol that the news articles belong to. The second parameter needs to implement IAsyncServiceRequest because this controls how the callback should be executed and provides the mechanism for intercepting the callback. The following code sample shows the signature for the BeginGetNews method. This method was made virtual, because the interception mechanism uses the VirtualMethodInterceptor.

C#
```
public virtual void BeginGetNews(string tickerSymbol,
   AsyncServiceRequest<ServiceListResult<NewsArticle>>
   asyncRequest)
```

The AsyncMethodCachingCallHandler checks if the requested news articles are already stored in the cache. If they are present in the cache, the cached news articles are returned immediately, by executing the callback. If the values were not in the cache, then the Async MethodCachingCallHandler will register itself to intercept the callback when it is executed. Because this is a callback, this cannot be done using Unity interception, so the callback will get intercepted explicitly by the AsyncMethodCachingCallHandler. The following code shows how the AsyncMethodCachingCallHandler intercepts the request:

C#
```
public IMethodReturn Invoke(IMethodInvocation input,
   GetNextHandlerDelegate getNext)
   {
      // The first parameter should identify the data in the
      // cache. The second parameter should be an
      // IAsyncServiceREquest
      IAsyncServiceRequest request = input.Arguments[1] as
         IAsync-ServiceRequest;
      if (request == null)
      {
         throw new InvalidOperationException(
            "Second parameter is not assignable to
            IAsyncService-Request");
      }
   // See if the item was already cached.
```

```
    var cachedValue = (objectCache.Get(BuildCacheKey(input)));
    if (cachedValue != null)
    {
        // Item was already cached. Raise the callback with
        // the cached value and return immediately
        request.ReturnResult(cachedValue);
        // By returning here, we can prevent the intercepted
        // object from being called
        return input.CreateMethodReturn(null);
    }
    // The item was not already cached. We'll need to wait
    // until the callback and then return
    request.OnCallback += (sender, args) =>
        objectCache.Set(BuildCacheKey(input), args.Result,
        policy);
    // continue with calling the original intercepted method
    return getNext()(input, getNext);
}
```

Once the news articles have been loaded by the News Service Agent and it executes the callback to notify the UI, this method also gets intercepted. The AsyncMethodCachingCallHandler will then add the loaded news articles to the cache.

The Stock Trader V2 automatically sets up all types for interception through policy injection using the VirtualMethodInterceptor. How this works is described in more detail in the section called **Enabling Policy Injection for all Registered Types**. The NewsModule, which is also responsible for registering the NewsServiceAgent to the container, also creates a policy that sets up the AsyncMethod CachingCallHandler. The following code shows how this works:

C#

```
unityContainer.Configure<Interception>()
    .AddPolicy("Cache")
    .AddMatchingRule<TypeMatchingRule>(
    new InjectionConstructor(
    new InjectionParameter(typeof(NewsServiceAgent))))
    .AddCallHandler<AsyncMethodCachingCallHandler>(
    new InjectionConstructor(new ResolvedParameter(typeof
    (ObjectCache)), new InjectionParameter(
    cacheItemPolicy)));
```

Catching and Handling Exceptions from the Service Boundary

The Stock Trader V2 relies heavily on web services for its information. While these web services are built in different technologies, such as regular Windows® Communication Foundation (WCF) Services, WCF RIA Services, and WCF Data Services, there are some common requirements with regards to exception handling:

- If a technical exception has occurred while accessing a service, we need to create a log entry for this exception so that the problem can be analyzed by the support organization.
- If the exception is a functional exception, for example a validation exception, then the validation results must be extracted from the exception and returned to the user interface.
- In the news module, exception handling for service requests is implemented using interception. With synchronous exception handling, this would be really easy to implement, because Enterprise Library ships with the ExceptionCallHandler, which is a policy injection call handler that you can use to catch and handle any exceptions that occur within an intercepted member (and that are surfaced by the method).
- With asynchronous methods, this technique cannot be easily applied. As described in the chapter on exception handling, if you wish to pass an exception back up to the user interface, you will need to do it explicitly.
- The following code shows the **OnNewsLoaded** method from the **NewsServiceAgent** class that's called when the news is loaded. If the **loadOperation** has an error, it will be returned to the user interface by calling the **ReturnError** method from the **AsyncRequest** class. The AsyncExceptionHandlingCallHandler class will intercept the exception and use the Enterprise Library Exception Handling Application Block to handle it. If the exception was not handled, then the **MarkErrorAsHandled** method will not be called. This will re-throw the exception, which will trigger the **Application.Unhandled_Exception** event.

C#

```
private void OnNewsLoaded(LoadOperation<NewsItem> loadOperation)
{
   var asyncRequest = AsyncServiceRequest<ServiceListResult<News
      Article>>.RestoreFromUserState(loadOperation.UserState);
   var firstEntity = loadOperation.Entities.FirstOrDefault();
   if (loadOperation.HasError)
   {
      // Return the exception to the user interface. Here, it
      // will be handled by the AsyncexceptionHandlingCallHandler
      // class.
      if (asyncRequest.ReturnError(loadOperation.Error).
         IsHandled)
      {
         loadOperation.MarkErrorAsHandled();
      }
      return;
   }
   if (firstEntity == null)
   {
      asyncRequest.ReturnEmpty();
      return;
   }
   List<NewsArticle> result = GetNewsArticles(loadOperation);
   asyncRequest.ReturnResult(
      new ServiceListResult<NewsArticle>(result));
}
```

10 Applying Interception

In this chapter, we'll discuss various ways of applying interception in an application.

If you wish to use interception on your objects, you'll need to take the following steps:

- Set up Unity to use interception
- Add the type registration
- Add the interceptor you'd like to use to the registration
- Add the behaviors that you'd like to use to the registration

Setting Up the Container for Interception

Before you can use interception, you should set up the unity container to use interception. The following example demonstrates how you can do this. If you are using the **EnterpriseLibraryContainer** with the default configuration, you don't have to do this because it will already have interception enabled.

C#

```
var container = new UnityContainer();
container.AddNewExtension<Interception>();
```

Intercepting Objects

As mentioned before, there are two different types of interceptors that you can use in Silverlight to intercept calls to an object: the **InterfaceInterceptor** and the **VirtualMethodInterceptor**. The **TransparentProxyInterceptor** is not available in Silverlight.

Intercepting an Object Using an Interface

It is possible to use interfaces for interception. The interception process will create a dynamic proxy that implements the requested interface. The following example shows how you could set up the **AccountHolderServiceAgent** for interception through the **IAccount HolderServiceAgent** interface.

```
C#
public interface IAccountHolderService
{
   void Update(AccountHolder accountholder);
}
public class AccountHolderServiceAgent :
   IAccountHolderService-Agent
{
   public void Update(AccountHolder accountholder)
   { // Implementation removed..
   }
}
public void SetupInterception(UnityContainer container)
{
   // Register the type mapping
   container.RegisterType<IAccountHolderServiceAgent,
      AccountHolderServiceAgent>(
      new InterfaceInterceptor(), // Add the type of interceptor
      new CustomInterceptionBehavior()); // Add the required
                                          // behavior(s)
}
public void GetInterceptedType(UnityContainer container)
{
   // When resolving, you'll get a dynamic proxy that wraps the
   // real Account Holder Service and intercepts any calls to it.
   var accountHolderService =
      container.Resolve<IAccountHolderService>();
   // This member call will be intercepted:
      accountHolderService.Update();
}
```

Interception Using Virtual Methods

You can also set up interception using virtual methods. The interception process will dynamically create a class that derives from the interception target and intercept all calls to virtual members. The following example demonstrates how you can use the **VirtualMethod Interceptor** to intercept calls to the **NewsServiceAgent**. The call to the BeginGetNews method will be intercepted.

```
C#
public class NewsServiceAgent : Infrastructure.Interfaces.
   INewsServiceAgent
{
   public virtual void BeginGetNews(string tickerSymbol,
   AsyncServiceRequest<ServiceListResult<NewsArticle>>
      asyncRequest)
   {
      // .. Implementation removed
   }
}
public void SetupInterception(UnityContainer container)
{
   // Register the type mapping
   container.RegisterType<INewsServiceAgent,
      NewsServiceAgent>(new VirtualMethodInterceptor(),
   // Add the type of interceptor
   new CustomInterceptionBehavior());
   // Add the required behavior(s)
}
public void GetInterceptedType(UnityContainer container)
{
   // When resolving, you'll get a dynamic proxy that
   // derives from NewsServiceAgent and intercepts calls to
   // all members that have been marked virtual
   var accountHolderService = container.
      Resolve<IAccountHolderService>();
   // This member call will be intercepted:
   Action<ServiceListResult<NewsArticle>> callBack =
      OnArticles-Loaded;
   accountHolderService.BeginGetNews ("Stock1", callBack);
}
```

Adding Interception to Already Registered Types

The previous examples demonstrate how you can set up an object for interception at the time the type is registered to the container. Sometimes you might also want to register interception after a type has already been registered. If you want to do this, you'll have to keep in mind that you need to add the interception to the proxy type. So if you are using interface interception, then you must register the interception against the interface, but if you are using virtual method interception, you must register the interception against the physical type.

The following code sample shows how you can do this:

```
C#
// set up interface interception for the IAccountHolderService-
// Agent interface, after it has already been registered.
// Note that we have to register the interception against the
// interface
container.RegisterType<IAccountHolderServiceAgent,
  AccountHolderServiceAgent>(
  new InterfaceInterceptor(), // Add the type of interceptor
  new CustomInterceptionBehavior()); // Add the required
                                     // behavior(s)
// Set up virtual method interception for the news service
// agent after it has already been registered.
// Note that we have to register the interception against the
// concrete type.
container.RegisterType<NewsServiceAgent>(
  new VirtualMethodInterceptor(), // Add the type of interceptor
  new CustomInterceptionBehavior()); // Add the required
                                     // behavior(s)
```

Adding Behavior to Intercepted Objects

Once you have set up an object for interception, you can add behaviors to the intercepted objects. Enterprise Library does not provide any default behaviors, but it does provide several policy injection call handlers to use with Unity's Policy Injection interception behavior. In this section we'll look at how you can implement a custom behavior and how you can implement additional interfaces on objects using interception.

Creating a Custom Interception Behavior

To create a custom interception behavior, you will need to create a class that derives from IInterceptionBehavior. The following code demonstrates how to intercept a call to the MyMethod method. Depending on the value of firstParameter, it will continue execution, return a different value, or throw an exception.

C#

```
public class MyClassToIntercept
{
   public virtual int MyMethod(string firstParameter, int
      second-Parameter)
   {
      return 0;
   }
}
public class ExampleInterceptionBehavior : IInterceptionBehavior
{
   public IMethodReturn Invoke(IMethodInvocation input,
      GetNextInterceptionBehaviorDelegate getNext)
   {
      // Check which method to intercept. In this case, we're
      // testing for a specific method name
      switch(input.MethodBase.Name)
      {
         case "MyMethod":
            return InterceptMyMethod(input, getNext);
         break;
      }
      // If you do not wish to intercept, then execute this
      // statement to continue to call the real MyMethod.
      IMethodReturn methodReturn = getNext()(input, getNext);
      // If you wish, you can perform post processing logic.
      return methodReturn;
   }
   private IMethodReturn InterceptMyMethod(IMethodInvocation
      input, GetNextInterceptionBehaviorDelegate getNext)
   {
      // Get the values for the parameters. You can also
      // change the values
      string firstParameter = (string) input.Arguments[0];
      int secondParameter = (int) input.Arguments[1];
      // You can execute logic before or after the method is
      // executed
      if (firstParameter == "Continue")
      {
         // Continue execution. This will eventually call the
         // real MyMethod()
         IMethodReturn methodReturn = getNext()(input,
            get-Next);
```

```
            // If you want, you can change the return value or
            // execute logic after the method is executed
            if ((int)methodReturn.ReturnValue == -1)
            {
                methodReturn.ReturnValue = 199;
            }
            return methodReturn;
        }
        else if (firstParameter == "ReturnDifferent")
        {
            // Stop execution and return something different
            return input.CreateMethodReturn(99);
        }
        else if (firstParameter == "Exception")
        {
            // Throw an exception and prevent the execution
            // of the intercepted method
            return input.CreateExceptionMethodReturn(
                new Exception("Return this exception"));
        }
    }
    public IEnumerable<Type> GetRequiredInterfaces()
    {
        return Enumerable.Empty<Type>();
    }
    public bool WillExecute
    {
        get { return true; }
    }
}
```

Implementing Interfaces Using Interception

You can use Unity interception not only to intercept calls to existing methods, but also to add additional interfaces to an object. For example, if you have a model that does not implement INotifyProperty Changed, you can use an interception behavior to add INotifyProperty Changed support to it.

In the Stock Trader V2 Reference implementation, this is demonstrated using the NotifyPropertyChangedInterceptionBehavior class. This interception behavior will intercept all calls to the object, including calls that might not exist on the original object. If those calls belong to the INotifyPropertyChanged interface, it will be handled by the behavior.

The following code shows how you can add INotifyProperty Changed support to an object.

```
C#
public class Model
{
   public virtual string Name { get; set; }
}
// First, set up the model for interception and include the
// INotifyPropertyChangedBehavior.
var model = Intercept.ThroughProxyWithAdditionalInterfaces<Model>
   (new Model(),
   new VirtualMethodInterceptor(),
   new[] { new NotifyPropertyChangedInterceptionBehavior() },
   new[] { typeof(INotifyPropertyChanged) });
// Since the object doesn't implement the interface itself,
// you will need to cast it to the interface
var npc = model as INotifyPropertyChanged;
// You can then register for the changed event
npc.PropertyChanged += delegate { };
// If you change the object, then the PropertyChanged event will
// be triggered.
model.Name = "Sharon";
```

This approach works best if you use the **VirtualMethodInterceptor** and mark the properties you would like to add the **NotifyProperty Changed** behavior to as virtual. The interception will then even work if you change the property values from within the object itself.

Applying Policy Injection

As mentioned in previous chapters, policy injection is a mechanism for interception that allows for finer control over which methods should be intercepted.

Setting Up a Type Mapping in Unity for Policy Injection

Policy injection in Unity is implemented as an interception behavior. So if you wish to set up a type mapping for policy injection, you will need to register the PolicyInjectionBehavior and choose which interceptor you would like to use. The following example shows how to use the InterfaceInterceptor to intercept calls to the IModel interface and how to apply policy injection to it:

```
C#
unityContainer.RegisterType<IModel>(
    new InterceptionBehavior<PolicyInjectionBehavior>(),
    new Interceptor<InterfaceInterceptor>());
```

Defining Policies in Code

Once you have set up a type for policy injection, you will need to specify what you would like to intercept and what the interception should do. You can do this by specifying one or more policies. All the policies that match are applied to an intercepted object by the policy injection behavior. Each policy has matching rules, which define which types and methods should be intercepted, and call handlers, which add behavior to the intercepted member calls.

The following code sample demonstrates how to add an interception policy. It uses the TypeMatchingRule to apply policy injection to the NewsServiceAgent class. It then adds the AsyncMethodCaching CallHandler to the policy, which applies caching to all intercepted methods.

```
C#
unityContainer.Configure<Interception>()
   .AddPolicy("Cache")
   .AddMatchingRule<TypeMatchingRule>(new InjectionConstructor(
      new InjectionParameter(typeof (NewsServiceAgent))))
   .AddCallHandler<AsyncMethodCachingCallHandler>(
      new InjectionConstructor(new ResolvedParameter(typeof(
      ObjectCache)), new InjectionParameter(cacheItem Policy)));
```

Appendix

WCF Data Services Validation in Silverlight QuickStart

The WCF Data Services Validation QuickStart is a sample application that demonstrates the use of the Validation Application Block in order to perform cross-tier validation across a Microsoft® Silverlight® client that consumes an OData feed exposed by a server using Windows® Communication Foundation (WCF) Data Services. For more information on validation and the Validation Application Block, refer to Validation.

WCF Data Services is a component of the Microsoft .NET Framework that enables you to create services that use the Open Data Protocol (OData) to expose and consume data over the web or intranet by using the semantics of representational state transfer (REST).

In Silverlight, you can also consume OData services. The Silverlight WCF Data Services components can generate a proxy for you that will contain a client-side representation of the business entities exposed by the OData service.

WCF Data Services doesn't have built-in support for validation, like WCF RIA Services, but you can use Enterprise Library to implement validation on the Silverlight client or the server (or both).

The QuickStart is based on the **WCF Data Services sample for Silverlight available on MSDN** and enhanced to support validation. You should use the QuickStart sample for educational purposes, but be aware that the scenario is not complete for production environments. The infrastructure pieces provided by the sample are neither complete nor properly tested, compared to the Validation Application Block, which is production-ready.

The sample shows one possible approach to accomplishing the cross-tier validation goal: the use of linked source code files in Microsoft Visual Studio® development system and the MetadataType attribute. There are other simple approaches to this challenge as well. For more information, see Performing Cross-Tier Validation Using the Validation Application Block.

Using the Metadata Type to Define Validation Rules

By default, both on the Silverlight client when using WCF Data Services tooling and on the server, if you use Entity Framework, the Visual Studio tooling will generate the code for your business entities. This means that you cannot add validation attributes to the properties of your business entities directly, but must add them to metadata classes that you have defined using the **MetadataType** attribute. This attribute is available in the .NET Framework, but it is not available out of the box in Silverlight 4. Because this attribute is critical for these kinds of situations, the Validation Application Block provides an implementation of this attribute.

> **Note:** *The new MetadataType attribute for Silverlight will be honored when performing validation using the Validator class included in the Validation Application Block, but will be ignored if using the Validator class from the System.ComponentModel.DataAnnotations namespace.*

The usage of this attribute enables a metadata class in Silverlight to externally annotate an entity class with validation rules, making this approach valuable not only to WCF Data Services, but also to any technology that automatically generates the entities with a tool. An example of a similar scenario where the MetadataType attribute is valuable is when using plain WCF services and data contracts.

The following code sample from the QuickStart demonstrates how you can add validation rules to the Order_Detail class in metadata:

```
C#
using System;
using System.ComponentModel.DataAnnotations;
#if SILVERLIGHT
using Microsoft.Practices.EnterpriseLibrary.Validation
   .DataAnnotations;
#endif

#if !SILVERLIGHT
namespace NorthwindDataService
#else
namespace SilverlightClient.Northwind
#endif
```

```
{
   [MetadataType(typeof(Order_DetailMetadata))]
   partial class Order_Detail
   {
      internal class Order_DetailMetadata
      {
         // using standard Data-Annotations attributes
         [Range(1, 50)]
         Int16 Quantity { get; set; }
         [Range(0, 1)]
         Single Discount { get; set; }
      }
   }
}
```

In this case, the sample is using linked files in the project in order to share these validation rules between the Silverlight client and the server. In order for this to work, conditional compilation instructions are needed because the MetadataType attribute in Silverlight is located inside the Validation Application Block default namespace and not in the System.ComponentModel.DataAnnotations namespace.

Additionally, because the namespace where the generated entities reside in the Silverlight client is not the same as the one used by the Entity Framework, conditional compilation is also used to account for this difference.

As shown above, you can take advantage of the built-in DataAnnotations attributes to add validation rules to your entities, but you can also expand these with the attributes provided by the Validation Application Block, as seen in the Order class.

C#

```
[MetadataType(typeof(OrderMetadata))]
partial class Order
{
   internal class OrderMetadata
   {
      // using attributes provided by VAB
      [IgnoreNulls]
      [RelativeDateTimeValidator(1, DateTimeUnit.Day, Range-
         BoundaryType.Inclusive, 3, DateTimeUnit.Month,
         RangeBoundaryType.Inclusive, MessageTemplate = "A
         change in the order requires that the required date be
         no less than one day from now, and no more than 3
         months.")]
      DateTime? RequiredDate { get; set; }
      [IgnoreNulls]
```

```
        [RangeValidator(typeof(decimal), "0", RangeBoundaryType.
           Inclusive, "0", RangeBoundaryType.Ignore)]
        decimal? Freight { get; set; }
    }
}
```

Once the attributes are in place, you will need to explicitly perform validation on both the client and the server, as WCF Data Services will not do this automatically.

For a list of all the available validation attributes provided by Enterprise Library, see **Using Validation Block Attributes**.

Performing Validation on the Server

The following example shows the usage of a ValidationHelper class in the service that will use the Validator class to perform validation on the incoming entities. If validation errors are found, the helper will raise a DataServiceException with a predefined message structure, so that the Silverlight client can then distinguish a validation error from a different kind of error. This provides a nice experience for the user, indicating how to fix the validation error. Even if the Silverlight client also performs validation in real time, there are some validation rules that can only be performed on the server side.

Regardless of whether the Silverlight client validates the data in its tier, the server always has to revalidate every input and enforce all of the validation rules because the client could, maliciously or not, bypass the client validation rules. Or, because of the open nature of the OData protocol, there might be multiple clients (also called consumers) of the WCF Data Services or OData feed that might not share the validation rules with the server.

C#

```
public class Northwind : DataService<NorthwindEntities>
{
    // ...
    [ChangeInterceptor("Orders")]
    public void OnChangeOrder(Order order, UpdateOperations
        operations)
    {
        // this helper will validate the entity, and if it
        // fails, it will throw a DataServiceException with
        // the validation error details, so that they
        // can be interpreted on the client.
        ValidationHelper.ValidateOrThrow(order, operations);
    }
}
```

Performing Validation on the Client

On the client, you can use the annotations on your entities to provide real-time validation feedback; for example, when a property value changes. The sample shows a very basic infrastructure to support this kind of automatic real-time validation: using partial classes to extend the entities generated by WCF Data Services and implementing the INotifyDataErrorInfo interface.

Also, when the user is about to submit the changes, you can explicitly trigger validation in order to validate all the entities that are about to be sent to the server, as seen in the following code.

C#

```
private void OnSaveChangesClick(object sender, RoutedEventArgs e)
{
   var updatedEntitites = context.Entities
      .Where(x => x.State == EntityStates.Modified ||
      x.State == EntityStates.Added)
      .Select(x => x.Entity)
      .ToList();
   if (updatedEntitites.Count == 0)
   {
      MessageBox.Show("No changes to submit.");
      return;
   }
   var results = new ValidationResults();
   foreach (var entity in updatedEntitites)
   {
      results.AddAllResults(Validation.Validate(entity));
   }
   if (!results.IsValid)
   {
      // TODO: show error validation results in a better way.
      MessageBox.Show("Please fix the validation errors before
      submitting."
      + string.Concat(results.Select(x => Environment.
      NewLine + x.Message)));
   }
   else
   {
      try
      {
         context.BeginSaveChanges(SaveChangesOptions.Batch,
         OnChangesSaved, context);
      }
```

```
        catch (Exception ex)
        {
            ShowGenericErrorMessage(ex, "The changes could not be
                saved to the data service.");
        }
    }
}
```

Additionally, because the server also performs validation and has a well-known message format to notify the client of validation errors, the following code demonstrates how you can differentiate a validation error from, say, an exception raised by WCF Data Services due to a communication channel issue or internal server errors.

C#

```
private void OnChangesSaved(IAsyncResult result)
{
    Dispatcher.BeginInvoke(() =>
    {
        context = result.AsyncState as NorthwindEntities;
        try
        {
            // Complete the save changes operation and display
            // the response.
            DataServiceResponse response = context.
                EndSaveChanges(result);
            // ...
        }
        catch (Exception ex)
        {
            ValidationResults results;
            // This helper returns the validation results in case
            // the data service exception was caused by a
            // validation error.
            if (ValidationHelper.TryGetValidationResults(ex, out
                results))
            {
                // Can store this in the validation summary
                // instead of showing a message box.
                MessageBox.Show(
                    string.Concat(results.Select(x => x.Message +
                    Environment.NewLine)),
                    "Validation Error",
                    MessageBoxButton.OK);
```

```
            }
            else
            {
                // Display the generic error (not due to validation).
                    ShowGenericErrorMessage(ex);
            }
        }
    });
}
```

Because you can submit multiple changes to different entities in one batch, you should find a way to correlate an error message with the entity that originated the error, especially when the end user would not be able to easily identify where to fix the errors. This scenario is not shown in the QuickStart code.

Prerequisites

To compile and run the WCF Data Services for the Silverlight Quick-Start, you will need the following prerequisites:

- Visual Studio 2010
- Microsoft SQL Server® 2008 Express edition (instance running on **localhost/SQLEXPRESS**)

Running the QuickStart

After you open the QuickStart solution, please ensure that the **NorthwindDataService** project is set as the startup project, with Default.html as the startup page.

Press F5 to run the QuickStart.

Index

A

.NET assembly, 41-42
.NET Framework, 43
AccountHolder class, 127-128
Account Holder Info Module, 130-132
AccountHolderService, 128
AccountHolderServiceAgent, 192
account information, 25
account position screen, 22
advanced (view) model scenarios, 136-145
aliases, 42-43
applications
  configuration console, 34-36
  implementing caching in, 163-170
  sample, xiii, xvii
  see also Silverlight applications
Application_UnhandledException event, 98-100
architecture, 27-28
  configuration, 48-50
asynchronous operations, 104-105
asynchronous service agent, 105-107
asynchronous web services, 93-94
AsyncMethodCachingCallHandler, 186-187
AsyncServiceRequest class, 100-101
AsyncXAMLConfigurationMerger, 50
atomic operations, 153-154
attributes, 112-114
audience, xiv

B

BeginGetNews method, 186
behaviors, 194-197
browsers
  cache circumvention, 47
  caching, 149
  log service, 74-75
buffering to isolated storage, 63
business entities, 118-119

C

CacheItemPriorities, 150-153
caches
  capabilities, 153-154
  expiration, 149
  item policy, 152
caching, 8, 147-170
  implementing in application, 163-170
  proactive, 159-161
  Stock Trader V2, 157-163
  values, 159
Caching Application Block, 147-148, 163-167
  see also Enterprise Library Caching Application Block
Caching track, xvi
call handlers, 183
Chain of Responsibility pattern, 179-180
clients, 203-205
code creation with the wizard, 37-38
command line, 39-40
components, 15-16

configuration, 31-44
    adding to an already existing server, 47
    architecture, 48-50
    Caching Application Block, 163-167
    Enterprise Library, xvi, 31-32
    files, 48
    on server, 75-76
    sharing validation rules, 120-121
    on Silverlight, 77-78
    support, 5
configuration console
    applications, 34-36
    installing, 14-15
    selecting types in, 35-36
    Visual Studio integration, 34-35
    XAML configuration files, 36-37
containers, 191
Contains method, 169, 176
contributors *see* personas
cross cutting concerns, 173-175
cross-tier validation, 116-117
current release, 3
custom exception handler, 107-108
custom extensions, 40-42
custom interception behavior, 194-196

D
database scripts, 14
DataContractJsonSerializer, 156
DataServicesValidationHelper, 129
data transfer objects, 116-119
default code generator, 123-124
derived class, 80-81
design
    Logging Application Block, 56-59
    Silverlight Caching Application Block, 151
developers, 70-73
diagnostic logging and tracing, 51-90
    Enterprise Library, 51-54
    Silverlight applications, 53-54
diagnostics
    information, 26-27
    logging and tracing, 73-90
    Logging Application Block, 55
    which are not in Silverlight, 54
documentation, 17

E
EditAccountHolderView, 145
Ed the Enterprise Library expert, xviii
Enterprise Library, xiv, 3-4
    adding configuration to an already existing server, 47
    configuring, 31-32
    custom extensions, 40-42
    diagnostic logging and tracing, 51-54
    installing components, 13-14
    Integration Pack, 3
Enterprise Library 5.0
    changes, 115-116
    earlier versions of, 16
Enterprise Library Caching Application Block, 150
EnterpriseLibraryContainer, 191
EnterpriseLibraryServerValidator class, 126
Enterprise Library Silverlight Integration Pack, xvi
    installing, 10
Enterprise Library Validation Application Block, 4, 111-129
    cross-tier validation, 116-117
    validation, 122-123
    WCF Data Services, 127
    WCF RIA Services, 114-115, 121-126
errors
    centralized viewing of, 68-69
    detecting, 53
    displaying messages, 107-108
    not bound in the UI, 144-145
ErrorsCollection class, 137, 140
exception handling, 91-108
    asynchronous web services, 93-94
    custom, 107-108
    exceptions, 105-105
    global exception handler, 92-93
    policies, 96
    Service Boundary, 188-189
    Silverlight, 92-95
    Stock Trader V2 RI, 97-108
    when calling services, 100
Exception Handling Application Block, 5, 91-92
    see also Silverlight Exception Handling Application Block
ExceptionManager class, 91-92
Exception Manager diagram, 99
Export to XAML wizard, 9
extended properties, 59
Extension Manager, 10
    installation, 11-13
extensions, 40-42

G

global exception handler, 90-101
guidance module, 24

H

handlers, 94
how to use this guide, xv

I

ICallHandler interface, 181
IInterceptionBehavior interface, 177-178
ILogEntryRepository interface, 63
IModel interface, 195-196
InMemoryCache, 152
in-memory cache, 161-162
In-Memory Caching scenario, 4
INotifyDataErrorInfo interface, 127, 138-139
INotifyPropertyChanged interface, 127, 136-138, 195
installation
    configuration console, 12-13
    Enterprise Library Silverlight Integration Pack, 8
    learning materials, 15
    NuGet package, 9-11
instances
    intercepting, 173-175
    multiple, 154-155
    see also interception
instrumentation, 57
Integration Pack, 1-2
interception, xvi, 6, 169-181
    applying, 189-196
    behaviors, 177-178
    containers, 189
    custom behavior, 192-194
    implementation interfaces, 194-195
    intercepted objects, 192-195
    for interfaces, 190
    objects, 189-192
    policy injection, 178-181
    registered types, 191-192
    virtual methods, 190-191
    see also News Module
InterfaceInterceptor, 174, 189
interfaces
    for interception, 190
    interception implementation, 194-195
isolated storage, 63
    asking for more space in, 159-161
    buffering to, 61
    caching values, 157
isolated storage cache, 152-155
    adding, 163-165
    multiple instances of, 154-155
Isolated Storage Caching scenario, 4
IsolatedStorage class, 159
Isolated Storage Log Entry Repository, 82
Isolated Storage Trace Listener, 50, 62-63, 79-81
IT *see* Mark the IT pro responsible for maintenance
item size on disk, 153

J

JSON serializer, 154-155

L

learning materials, 15
levels, 85-86
linked files, 115-116
Log Entry objects, 56
log filters, 56
log formatters, 57
logging, 5
    changing for trace source, 85-86
    correlating messages with activities, 69-71
    diagnostic, 71-88
    endpoints, 61-62
    handlers, 94
    increasing levels at run time, 68
    Stock Trader V2, 65-71
    toggling, 85
    see also diagnostic logging and tracing
Logging Application Block
    changing levels, 85-86
    configuring on server, 73-74
    design of, 54-57
    diagnostics, 53
    features not ported, 57
    Remote Service trace listener, 76-77
    XAML configuration files, 30-31
    see also Silverlight Logging Application Block
Logging Service Factory, 60, 78
Logging track, xvi
logs
    adding services to browser, 72-73
    customizing services with a derived class, 78-79
    messages, 82
Log Writer object, 55

M

maintenance, xx
map of this book, xv
mapping
    .NET Framework and Silverlight, 43
    policy injection, 197-198
market trend analysis, 22-23
Mark the IT pro responsible for maintenance, xx
matching rules, 181-182
maximum size, 65
memory, 148
MemoryBackedCache class, 151-152
messages
    with activities, 71-73
    buffering, 63
MetadataTypeAttribute, 124
MetadataType attribute, 127-128, 200-202
Model Graph Validator, 143
ModelGraphValidator class, 137, 141-143
ModelSupport class, 138
ModelSupport library, 136-138
Model View ViewModel pattern, 28
modular development, 29-30
MSBuild task, 38-39
MSI, 15-16
multi-threading support, 145
MyMethod method, 194-196

N

.NET assembly, 41-42
.NET Framework, 43
new keyword, 177
news articles, 23-24
News Module, 185-189
non-default configuration sources, 36
notification events, 86-87
Notification Trace Listener, 65, 84-86
NotifyingModel class, 138, 143
NuGet package, 3, 9
    installation, 11-13

O

ObjectCache class, 151-152, 167-170
    adding items to, 168
    RemovedCallback, 169-170
    retrieving from container, 167
    testing for presence of items, 168-169
ObjectCollectionValidatorattribute, 124-125
objects, 191-197
ObjectValidatorAttribute, 124-125
OnNewsLoaded method, 188-189
Open Data Protocol (OData), 127, 199-200
Optional Update xiii, 8
Order_Detail class, 200-201
OrderListViewModel class, 103
Orders module, xix, 132-135

P

partial classes, 120
personas, xiii, xvii-xx
policies
    defining in code, 198
    injection, 8, 197-198
    injection and interception, 180-183
position module, 22, 157-163
power users, 69-70
prerequisites, 10-11
    WCF Data Services Validation QuickStart, 205
Prism, 19
proactive caching, 159-161
properties, 59
PropertyValue class, 137-140

Q

QuickStarts, 17

R

RangeValidatorAttribute, 8
real-time validation, 203-205
registered types, 193-194
relevant news articles, 23-24
Remote Service trace listener, 60-61, 73-81
    changing, 89-90
    Logging Application Block, 78-79
    parameters, 61
RemovedCallback, 169-170
replace handler, 96
requirements, 10-11
Revert Impersonation property, 59
Runtime Changes API, 65

S

sample application, xiii, xvii
scenarios
    advanced (view) model, 136-145
    supported, 3-5
    Woodgrove Bank, 19
scripts, 14
security considerations, 44
servers, 46
    on validation, 202
server-side validation, 125-126, 128-129
service agent, 105-107
ServiceAgent class, 132
Service Boundary, 188-189
ServiceListResult class, 101-102
ServiceResult class, 102
services, 100-103
    exception handling, 100
sharing validation rules, 120-121
Sharon the Silverlight developer, xix
Silverlight
    caching considerations, 148-150
    exception handling, 92-95
    Integration Pack, xxii-xxiii
    mapping, 43
    some diagnostic information not available in, 54
    web service calls, 54
Silverlight 4 and Silverlight 5, 53
Silverlight applications
    consuming XAML configuration, 45-47
    in validation, 109-111
Silverlight Caching Application Block, 150-157
    design, 151
    Enterprise Library Caching Application Block, 150
    scenarios, 6
Silverlight Exception Handling Application Block, 94-96
    exception handlers, 96
Silverlight Logging Application Block, 7
    see also Logging Application Block
source code, 17
stack traces, 59-60
stand-alone configuration console, 35
stock trade orders, 24-25
StockTraderRI.Infrastructure.ModelSupport library, 136-138
Stock Trader V1, 19-20
Stock Trader V2, xiii
    caching, 157-163
    configuration files, 48
    logging and tracing, 67-73
    overview, 21-28
    validation, 129-145
    see also News Module
Stock Trader V2 reference implementation, overview, xvi
Stock Trader V2 RI, 19-30
    diagram, 30
    downloading, 17
    exception handling, 97-108
storage size limit, 147
supported scenarios, xxiii-xxvi
supported validators, 113-114
system requirements, 10-11
System.Runtime.Caching namespace, 150-151

T

TargetObject, 172
thread safety, 153-154
toggling, 87
ToString method, 8
trace listeners, 58
    adding to trace sources, 88
trace sources, 58
    logging, 87-88
    trace listeners, 88
tracing
    diagnostic, 73-90
    exposing to developers, 70-73
    see also diagnostic logging and tracing
tracks, xvi
TransparentProxyInterceptor, 8, 176-177, 191
TypeMatchingRule, 198
types
    intercepting, 177-179
    mapping between .NET Framework and Silverlight, 43
    mapping for policy injection, 197-198
    registered, 193-194
    selecting in configuration console, 35-36
    Silverlight 4 and Silverlight 5, 43

U

unhandled exceptions, 99-100
Unity Interception 2.1, 9
Unity interception for Silverlight, 175-180
usage patterns, 53

V

ValidatingModel class, 138
ValidatingModelSupport class, 138
validation, 6, 109-145
- in a Silverlight application, 109-111
- attributes, 8
- on the client, 203-205
- Enterprise Library Validation Application Block, 122-123
- errors not bound in the UI, 144-145
- MetadataType attribute, 200-202
- real-time, 203-205
- on the server, 202
- server-side, 125-126, 128-129
- sharing validation rules, 120-121
- Stock Trader V2, 129-125
- supported validators, 113-114
- WCF RIA Services, 122
- when property changes, 144
- *see also* Enterprise Library Validation Application Block; server-side validation

ValidationHelper class, 202
validation rules
- defining with attributes, 112-114
- linked files, 117-118

ValidationSummary control, 144
Validation track, xvi
version numbers, 8
VirtualMethodInterceptor, 201
virtual methods, 192-193
Visual Studio integration, 34-35

W

Warnings rule set, 133
WCF, 118-119
WCF binding
- configuring on Silverlight, 77-78
- Logging Service Factory, 80

WCF Data Services, 6
- Account Holder Info Module, 130-132
- Enterprise Library Validation Application Block, 127
- sharing validation rules, 120-121

WCF Data Services Validation QuickStart, 199-205
- prerequisites, 205

WCF RIA Services, xix, 6
- default code generator, 123-124
- Enterprise Library Validation Application Block, 114-115, 121-126
- Model Graph Validator, 143
- **Orders** module, 132-135
- validation, 122

web services, 93-94
- calls, 54

Windows Event Log, 54
Woodgrove Bank scenario, 19
wrap handlers, 96

X

XAML
- aliases, 42-43
- translation considerations, 43-44

XAML configuration files, 32-34
- circumventing the browser cache, 47
- code creating with the wizard, 37-38
- command line, 39-40
- configuration console, 36-38
- consuming from a Silverlight application, 45-47
- embedding in the XAP, 45
- loading from the server, 46
- Logging Application Block, 32-33
- MSBuild task, 38-39
- security considerations, 44

XAP, 45

Printed in Dunstable, United Kingdom

91343487R00132